# Project Management for IT–Related Projects

Textbook for the ISEB Foundation Certificate in
IS Project Management

## The British Computer Society

The British Computer Society is the industry body for IT professionals and a chartered engineering institution for information technology (IT). With members in over 100 countries, the BCS is the professional and learned Society in the field of computers and information systems.

The BCS is responsible for setting standards for the IT profession. It is also leading the change in public perception and appreciation of the economic and social importance of professionally managed IT projects and programmes. In this capacity, the Society advises, informs and persuades industry and government on successful IT implementation.

IT is affecting every part of our lives and that is why the BCS is determined to promote IT as the profession of the 21st century.

## Joining the BCS

BCS qualifications, products and services are designed with your career plans in mind. Belonging to the BCS doesn't just give you value for money. It gives you benefits money just can't buy: provides industry-recognized endorsement of your credentials; underlines your professionalism; supports your career development at every stage; helps you get your next job or contract; gives you access to an unrivalled network of IT professionals and groups; and provides a wealth of IT industry information.

Being a member of the BCS demonstrates your commitment to both your own and the IT community's professional development. With employers and customers increasingly requiring proof of professionalism – in terms of code of conduct, qualifications and competence – your BCS membership helps set you apart from other IT practitioners, confirming your status as an IT practitioner of the highest integrity.

BCS membership is also your pathway to Chartered Status, with the new post-nominals CITP (Chartered IT Professional).

www.bcs.org/membership

## Further Information

Further information about the British Computer Society can be obtained from: The British Computer Society, 1 Sanford Street, Swindon, Wiltshire, SN1 1HJ.

Telephone: +44 (0)1793 417424

Email: bcshq@hq.bcs.org.uk

Website: www.bcs.org

# Project Management for IT-Related Projects

Textbook for the ISEB Foundation Certificate in
IS Project Management

BOB HUGHES, ROGER IRELAND, BRIAN WEST,
NORMAN SMITH and DAVID I. SHEPHERD

THE BRITISH COMPUTER SOCIETY

**The British Computer Society,**
**1 Sanford Street,**
**Swindon, Wiltshire SN1 1HJ, UK**
**www.bcs.org**

ISBN 1-902505-58-1

British Cataloguing in Publication Data.
A CIP catalogue record for this book is available at the British Library.

Designed and typeset by Pete Russell, Faringdon, Oxon
Printed by Biddles Ltd., Kings Lynn. www.biddles.co.uk

# Contents

## 6 Estimating 83

*Learning outcomes* 83

- **6.1** Introduction 83
- **6.2** What we estimate and why it is important 83
- **6.3** Expert judgement 84
- **6.4** Approaches to estimating 85
- **6.5** A parametric approach 87
- **6.6** Estimating by analogy 90
- **6.7** Checklist 91

*Sample questions* 91
*Answers to sample questions* 92
*Pointers for activities* 92

## 7 Risk 95

*Learning outcomes* 95

- **7.1** Introduction 95
- **7.2** Risk management 96
- **7.3** Identifying risks 97
- **7.4** Assessing the risk 98
- **7.5** Deciding the appropriate actions 101
- **7.6** Planning, monitoring and control 103
- **7.7** Summary 104

*Sample questions* 105
*Answers to sample questions* 105
*Pointers for activities* 105

## 8 Project Organization 107

*Learning outcomes* 107

- **8.1** Introduction 107
- **8.2** Programmes and projects 107
- **8.3** Identifying stakeholders and their concerns 108
- **8.4** The organizational framework 108
- **8.5** Desirable characteristics of a project manager 113
- **8.6** Project support office 114
- **8.7** Project team 116
- **8.8** Matrix management 116
- **8.9** Team building 118
- **8.10** Team dynamics 119
- **8.11** Management styles 120
- **8.12** Communication methods 121

*Sample questions* 122
*Answers to sample questions* 122
*Pointers to activities* 122

## *About ISEB* 125

## *Index* 127

# Contributors

**BOB HUGHES** is an examiner and a former chief moderator for the ISEB Foundation Certificate in IS Project Management and is also an examiner for the BCS Professional Examination. Before becoming an academic, Bob Hughes had many years of experience of implementing IT projects in the telecommunications, energy and government sectors.

**DAVID I. SHEPHERD** has worked within the IS/IT industry since 1966 as computer hardware engineer, software programmer, project manager and lecturer. He has experience of developing a wide variety of computer systems including telecommunications systems, naval weapon systems, computer aided design systems and transaction processing systems. He has worked for organizations including the Ministry of Defence, Vosper-Thorneycroft, Racal, Philips BV, North East Worcestershire College and Quest Automation. He is currently the Academic Director at the Royal Military College of Science, responsible for both short courses and a Master of Science course covering IT project management and systems development.

**ROGER IRELAND** graduated from London University with a degree in electrical engineering, before working at Government Communications Headquarters (GCHQ) in Cheltenham. After a stint lecturing in further education, including teaching maths and computer science at the North Gloucester Technical College, Roger joined the instructor branch of the Royal Navy. His naval service included appointments in meteorology, oceanography, and IT development in the Joint Services IT Training Headquarters. Since his retirement, he has been heavily involved in ISEB activities including positions as an examiner and chief moderator.

**NORMAN SMITH**, who is currently self-employed as a tutor in project management, has spent more than thirty years in the IT profession, as a practitioner, consultant and trainer in Project Management and Systems Analysis. He has worked in senior roles for various international blue chip companies including Shell and the Royal Sun Alliance Insurance Group. Norman holds a BSc in Computer Science from Heriot Watt University, and has lectured at the Civil Service College. He is also an examiner for the Institute for the Management of Information Systems (IMIS), the Association of Chartered Certified Accountants (ACCA) and ISEB.

**BRIAN WEST** began work as an analyst programmer working in Assembler and COBOL. As Computer Training manager for British Steel he was responsible for systems and software training. At the National Computer Centre he managed the development of its major training products covering Systems Analysis and Design, COBOL, SSADM (V3 & V4), Telecommunications Management and Project Management. As Quality Manager for NCC Education he was responsible for introducing ISO 9000 and managed their International Examinations programme. He is currently working as an independent consultant, an ISEB examiner and a marker for the Association of Chartered Certified Accountants (ACCA).

# Figures and tables

# Introduction

 This publication is designed to support candidates who wish to take the examination leading to the Information Systems Examination Board (ISEB) Foundation Certificate in Project Management for IT-related Projects.

Through the ISEB, the British Computer Society provides industry-recognized qualifications that measure competence, ability and performance in many areas of IT, with the aim of raising industry standards, promoting career development and providing a competitive edge for employers.

ISEB qualifications certify employees' skills and competencies. They provide the means and the platform for recognizing and enhancing development through the employees' professional careers.

The Foundation Certificate in IS Project Management was launched in 2004 by ISEB. Although there are no prerequisites to taking the certificate, it is strongly recommended that candidates attend a training course accredited by the IS Examinations Board in preparation for the one hour, 40 question multiple-choice examination. This text has been created to support these training courses and prepare candidates taking the subsequent exam.

The course, this text and the examination are all designed to ensure that the successful candidates have a basic grounding in IT project management. They have been designed not only for those with a technical background in IT who are now beginning to take on project management responsibilities but also for users or potential users of IT systems who are involved in the implementation of IT systems in their place of work. Anyone involved in or affected by IT projects could benefit from this material, including users, buyers and directors. In Chapter 1 of this text, for example, we describe the usual stages of the system development life cycle, with which technically-oriented readers may already be familiar, but which may be helpful to those from the user community.

The text will act as the first stepping stone in acquiring a more formal knowledge of proper IT project management principles for those about to take up a management role. Those needing to acquire and demonstrate a deeper and more practical expertise in IT project management should consider going on to the full ISEB Certificate in IS Project Management. However it is envisaged that the text will also be useful for both developers and users who are involved at a less prominent management level in IT projects, but still need to understand how IT projects should be organized.

The ISEB course has a focus that is different to that of courses in, for example, PRINCE. PRINCE is a UK government-sponsored set of procedures for managing major projects. In our view, PRINCE effectively describes an information system for a project that allows it to be run in a controlled and efficient manner. Although PRINCE will tell you what decisions need to be taken, when they need to be taken and by whom, it offers little guidance about how decisions are made. We hope that the ISEB course will help to fill this gap.

The ISEB course can also be distinguished from more general introductory courses on project management by its focus on IT projects. While the core elements of project management remain the same regardless of the type of project, there are some significant differences in emphasis with IT projects. The description of the IT-focused system development life cycle has already been mentioned but there are other topics like testing and the use of function points to estimate system size which get more attention here than in more general project management courses.

While candidates no doubt want to pass the ISEB Foundation Certificate examination, the point of the course and this text is to improve the management of IT projects, where there have been cases of huge losses caused by project failure. In some places, we go into more detail than is strictly needed for the examination, in order to give readers the practical understanding they need in a work situation.

The following people have contributed the material for the text:

| Norman Smith | Chapters 1 and 4 |
| Bob Hughes | Chapters 2 and 6 |
| Roger Ireland | Chapters 3 and part of 8 |
| Brian West | Chapters 5 and part of 8 |
| David I. Shepherd | Chapter 7 |

Any defects and errors are almost certainly those of the editor, Bob Hughes. Sue McNaughton and Elaine Boyes at the BCS have driven the publication project along. The development of the Foundation Certificate as a whole has involved many BCS staff including Malcolm Sillars, Rebecca Stoddart, Imelda Byrne, Steve Causer, and Carol Lewis. Many errors and omissions in the text have been detected by the eagle-eyed Shena Deuchars.

The book is dedicated to the memory of Jimmy Robertson.

# CHAPTER ONE

# Projects and project work

## LEARNING OUTCOMES

When you have completed this chapter you should be able to demonstrate an understanding of the following:

▶ the definition of a project;

▶ the purpose of project planning and control;

▶ the typical activities in a system development life cycle;

▶ system and project life cycles;

▶ variations on the conventional project life cycle;

▶ implementation strategies;

▶ the purpose and content of the business case;

▶ types of planning documents;

▶ post-implementation reviews.

## 1.1 Projects

In general, a **project** may be defined as a **group of related activities carried out to achieve a specific objective.** Examples of projects include building a bridge, making a film and re-organizing a company. In this text we focus on projects that implement new information technology (IT) applications within organizations. These are technical but also involve changing the organization in some way. Before a project starts one or more people will have an idea about a desirable product or change. The project that emerges from this idea should possess the following attributes:

● A defined start point, which is when:

  • the exploration of the idea is converted into an organized undertaking;

  • the idea obtains business backing and a **project sponsor** – an individual or group within the organization who will provide financial resources for the project;

  • a commitment is made to provide the necessary resources;

  • responsibilities are defined;

  • initial plans are produced;

● A set of objectives, which:

  • drive the actions of the project team towards achieving a common goal;

- should be stated and understood at the start of the project;
- should be clear and unambiguous;

- A set of outputs or deliverables, which allow the objectives to be satisfied;

- A defined end point at which the objectives have been met – unless the project has been abandoned;

- A unique purpose – repeated activities are not projects;

- Benefits for the organization for which the project is being carried out which justify the project:

  - are quantifiable, ideally;
  - outweigh the costs.

    If, for example, the project is demanded by legislation, the business benefits may not outweigh the costs.

## 1.2   Successful projects

To be successful, a project should meet the following criteria:

- It should enable the stated objectives to be achieved.

- It should be delivered on time.

- It should be delivered within budget.

- It should deliver a system that performs to agreed specifications, including those relating to quality.

- It should provide value for money – bearing in mind that different organizations will have different perceptions of what constitutes 'value'.

- It should satisfy the project sponsor and other interested parties.

In summary, this means developing the project at a **specified cost**, within a **specified time**, to meet a **specified business requirement**. These three specifications are closely linked and any change to one will affect the others.

The sponsor and users typically want a system capable of a multitude of functions, to be delivered immediately and at low cost. As a general rule not all of this can be delivered and so the agreed project objectives will be a compromise between mutually exclusive requirements.

If the requirements stray outside this area of compromise by increasing cost or time, or by reducing functionality, then the project could cease to be viable. This happens, for example, if the costs of the project exceed the value of its benefits. However there may be exceptional circumstances in which a project can, with the sponsor's agreement, fail to meet one or more of these success criteria and yet still be considered a success.

### Canal Dreams booking system project scenario

Canal Dreams is a major holiday company that specializes in canal holidays. The present organization is the result of the acquisition of six regional canal boat leasing operators, as Canal Dreams has, over a number of years, acquired a number of previously family-run boatyards.

Each of the constituent boatyards has operated its own paper-based booking system, but now an IT-based, 'one-stop' booking system is to be introduced. It will be physically situated centrally at the Canal Dreams head office in Manchester. The new system should allow potential customers to enquire about and book holidays based in any region through a single contact point. The local boatyards should also be able to deal with bookings for customers who drop in, by accessing the central booking system from workstations in the boatyard offices. They will also need access to the central booking details to book boats out to and in from clients at the start and end of their holidays. It should be noted that currently the head office has no staff who deal with bookings.

Although Canal Dreams has some IT staff providing user support for desktop applications, it has no in-house software development staff. The intention is to contract systems design and building to an external supplier.

The Canal Dreams project scenario will provide examples throughout the text. The main objective in this scenario is the implementation of a central booking system. This will have business benefits for Canal Dreams. With the new system, a potential customer will only ever have to phone one number to book a boating holiday: if a boat is not available at one yard, the booking clerk will be able to book a boat at some other Canal Dreams boatyard. At present the customer would have to make another phone call; at this point, some customers might decide to call another company altogether, so the new system should increase sales. The centralized booking system should also allow Canal Dreams to cut costs by reducing the number of staff who take bookings at local boatyards.

In order to meet these business objectives, the proposed system will need certain **functionality**. For instance, it should allow the operator to check the availability of a boat at a particular boatyard in a particular week. There will also be **quality requirements**: for example, the time it takes the computer to respond to a user query on boat availability needs to be quite short as the operator will probably be talking to a customer.

If the system were to exceed the **cost requirement**, the potential additional income through extra bookings and staff savings might not be enough to meet the cost of implementing the system. Canal holidays are a seasonal business and so the transfer to the new system will need to be at a quiet time of the year before the bookings for the next season start to come in. This implies a certain **deadline** for system implementation.

## 1.3  Project management

Having established and agreed objectives, how do we then achieve them? The first step is good planning. Having produced good plans, this should be followed by monitoring and effective control of the project to fulfil the plans. The plan and the ways in which we monitor progress must be focused on achieving the agreed objectives. Someone has to take responsibility for controlling the work in accordance with the plans. This is the role of the **project manager**.

Successful project management cannot be guaranteed but certain elements which contribute to success can be recognized.

- **Clearly defined responsibilities**: it is essential that roles and responsibilities be clearly defined, documented and agreed. Vaguely defined responsibilities can only cause problems.

- **Clear objectives and scope**: any manager who embarks upon a project without clearly establishing the scope of the expected deliverables, together with cost, time and quality objectives, is creating problems for the future.

- **Control**: despite their individual differences, all projects can be controlled. It is important to establish at the outset how best to control the work and how to exercise that control.

- **Change procedures**: ideally the project manager would like to work in a world where there is no change or uncertainty. Unfortunately it has to be recognized that there is uncertainty and that change will happen. Appropriate procedures should be put in place to deal with it.

- **Reporting and communication**: clear reporting of project progress and any problems allows action to be taken quickly to resolve problems. Effective communication with all stakeholders can help avoid conflicts.

A project management method is a set of processes used to run a project in a controlled and, therefore, predictable fashion. The design and development procedures by which the objectives of the project are satisfied – for example, the use of object-oriented analysis – constitute the development method.

There are various project management methods which complement the development methods that can be used. In general, project management methods are applicable to a range of project types whereas development methods tend to be specific to projects with particular types of deliverable or objective. This is because the development tasks will vary according to the objectives of the project. Organizing an office move, developing a software application and providing disaster recovery facilities are all projects in their own right. Each has a different method of development but all are controllable using the same project management processes. Following a method does not guarantee that a project will be successful. If applied carefully, however, it will provide management with the means to be successful.

Assume that you are a manager of an office department that is going to be relocated to a building five miles away. Day to day management of the move will be delegated to one of your staff. What would be the main sequence of activities needed to plan and carry out the move? You want to leave as much of the detailed work as possible to your subordinate, but at what key points would you need to be involved to check progress?

*Activity 1.1*

*Solution pointers for the activities can be found at the end of the chapter.*

## 1.4 System development life cycle

Breaking a development method or approach into a number of processes is a widely accepted practice. This allows systems to be designed and implemented using a methodical and logical approach. The precise number and names of these processes will vary from organization to organization. In some cases, stages will be combined or split. Generally speaking, the following processes belong to the **system development life cycle** (SDLC) that applies to IT projects:

- initiation;
- feasibility study;
- project set-up;
- requirements analysis and specification;
- design;
- construction;
- acceptance testing;
- implementation;
- maintenance.

Each process creates one or more tangible products or **deliverables**. An important aspect of this phased approach is that at the completion of each process, a formal decision should be made to accept the products created by the process. The delivery of the products of each process can act as a **milestone** at which we can judge the progress and continuing viability of the project.

There are several variations on this model (some of which are described later) as well as a wide variety of alternative deliverables. We describe below the typical content of each phase.

## 1.4.1 Initiation

The objective of project initiation is to decide the most appropriate way to handle a request for work, taking into account any business or technical strategies that the host organization might have.

It begins with recognition by the managers of an organization that it has a need that can only be satisfied by some form of project. The need might be a perceived problem to be solved, a request for something new in an existing system or the identification of some new way of delivering value to the organization. The initiation process checks that a problem or deficiency really exists and decides whether the proposed change appears to be desirable. This phase is typically short. The end result or deliverable is a decision by the project sponsor on whether to spend resources on further investigation of the feasibility of the proposal. **Terms of reference** should be drawn up, outlining the scope of the proposal to be investigated and authorizing staff to carry out the investigation. Staff carrying out the investigation need to have permission to gather information from those currently working in the areas affected by the possible new system, along with other stakeholders.

## 1.4.2 Feasibility study

The feasibility study assesses whether the proposed development is practical in terms of the balance of costs and benefits, the technical requirements and the organization's information system objectives. The deliverable is a feasibility report which, among other things, presents the business with a range of options aimed at providing a solution.

Three types of constraint influence the decision about the appropriate option:

- Technical constraints: can the proposed project be completed with the technology currently available?

- Time constraints: can the proposed project be completed in the available time?

- Budgetary constraints: are the necessary finances available from the business? It may be that the project can be completed on time to satisfy the user's requirements, but only at a prohibitive cost.

One or more of these constraints may prevent a project from being developed any further.

### 1.4.3    Project set-up

Based on the recommendation of the feasibility study report, the organization decides whether to go ahead with the full project. At this point a group variously called the steering committee, project board or project management board is set up to oversee the project in the organization's interests. A project manager needs to be appointed and an initial project team set up to start work.

At this stage more detailed planning for the project takes place. Terms of reference for the project to implement the new system, as opposed to simply investigating its feasibility, are drawn up.

### 1.4.4    Requirements analysis

This phase defines the requirements of the new system in detail and identifies each business transaction. The gathering of requirements will usually involve interviewing users. In some cases, mock-ups or **prototypes** of parts of the new system could be used to help the users clarify their ideas about the requirements. It is surprising how much additional information can be gleaned from spending just a little more time with the user.

At the end of this phase, some form of **requirements statement** is produced. This describes what the final system should be able to accomplish and lists all the major features of the end product. It forms the basis of the contract between the customer for the new system and the developers.

At this point an outline of the test data and expected results should be drafted. This is used to check that the delivered system conforms to the specification. The test cases will form the basis of acceptance testing (see Chapter 5).

 **What kinds of people should the business analyst interview in the Canal Dreams booking system project in order to obtain the requirements for the new system?**

*Activity 1.2*

### 1.4.5    Design

If it is decided to build a new system, rather than buying a ready-made or off-the-shelf application, then a design phase is needed. This activity translates the business specification for the automated parts of the system into a design specification of the computer processes and data stores that will be needed.

Where a new application is to be built, the elements to be designed include:

- inputs;
- outputs;
- processing;
- data and information structures.

Most software is designed by first determining the output of the computer application. The reasoning is that if you know the outputs then you can determine the inputs needed to create those outputs. Once the outputs and inputs are known, the processing which converts the inputs to outputs can be determined. The designer is also in a position to consider what information needs to be recorded and the structure this data needs to have.

The identification of the inputs, outputs, processing and information that the system will deal with is known as **logical design**. The **physical design** is concerned with the actual appearance of the input and output screens and the printed reports that will be produced by the implemented system. Several different physical designs could satisfy the same underlying logical design. Physical design in this phase is essentially concerned with the system as it will appear to the outside world. Further internal physical design of the internal software and data structures will take place in the next phase.

Where an off-the-shelf application is to be used, because it already exists, its design will already have been carried out. In this case, the problem is to find the package whose features most closely match the business requirements. A plan is needed for the process by which available packages are to be evaluated and the ones which represent good value selected. Evaluation may involve trying out demonstration versions of the software, site visits to existing users of the software, and the careful study of the suppliers' documentation. In some cases the existing software needs to be customized, that is, modified, to meet the organization's particular needs.

**List some of the input screens that the Canal Dreams booking system may need and which would need to be designed.**

*Activity 1.3*

## 1.4.6   Construction

This process has the objective of designing, coding and testing software and ensuring effective interaction between different transactions. For example, the Canal Dreams transactions include online recording of holiday bookings and batch processing, such as overnight printing of reminders for final payment. Procedure manuals will also be produced and new hardware may have to be acquired.

During this phase, the requirements statement is re-examined to ensure that it is being followed to the letter. Normally any deviations have to be approved through a formal change procedure (see Chapter 4).

## 1.4.7   Acceptance testing

When the construction of the system is complete, or an off-the-shelf package has been acquired, knowledgeable representatives of the user and IT support staff need to test it before its implementation as the operational system. It is inevitable that during this stage the user will uncover problems that the developer has been unable to detect.

This activity may overlap with implementation. For example, some tests need to be carried out on the system when it is actually installed on the equipment that will be used operationally.

### 1.4.8 Implementation

Here the project reaches fruition. Hardware that has been purchased is delivered and installed. Software is installed, users trained, and the initial content of databases set up (for example, in the case of Canal Dreams, details of the boats that can be hired will need to be input). There are various strategies for implementation which are discussed later in this chapter.

### 1.4.9   Review and maintenance

At an agreed interval after the system has been made operational, a **post-implementation review** – sometimes called a post-evaluation review – should be carried out by a business analyst who was not involved in the original project. The review checks that the operational system has actually delivered the benefits envisaged in the original feasibility report. Where some of the original objectives or requirements have not been met, changes may have to be made.

Changes are sometimes the result of the system not completely fulfilling its original requirements, but they may also be the result of users identifying new requirements. Changes may also come about due to government regulations or alterations in the way the organization does business. These changes can be made as maintenance work, or they can become projects in their own right.

## 1.5   Project management and the development life cycle

The processes described above represent groups of development activities which have to be performed to complete the project. There is also a need for management checkpoints at which the progress and direction of the project can be formally assessed.

Most projects contain elements of uncertainty which make it difficult to exercise control against precise targets. This uncertainty tends to be greatest at the beginning of the project, when little may be known about detailed requirements, but gradually decreases as the project progresses. This makes it difficult to plan the later phases of the project in detail at the beginning of the project. For example it would be difficult for Canal Dreams to plan in detail the installation of new IT equipment when it has not yet identified what types of equipment, if any, are needed at which boatyard.

For the purposes of control there is a need to break the project into manageable units of work. This is similar to dividing development into processes, except that it focuses on how best to manage the project. No two projects will be broken down identically because no single project structure can provide the best management control for all projects. In some cases, perhaps on smaller projects, two activities of the system development life cycle, such as design and construction, might be treated as one stage for management purposes. On larger projects a single activity of the development life cycle might be split into several management stages. For example the construction phase might be broken into different management stages dealing with the construction of different parts of the system.

The end of each stage is marked by a formal review which assesses the work done, the work to be done and whether the business case is still valid. The review concludes with a formal sign-off and approval to continue to the next stage of work by the managers or the development team and the project sponsors.

## 1.6   Elements of project management

The motivation behind consolidating or splitting up development processes is to ensure controllability of work. In addition, the project manager has to tailor project management procedures to minimize risks to the project and to maintain control over the project, while avoiding excessive management bureaucracy. Although the overall project management process remains the same, different projects require different levels of control. The amount of effort required for project management, for example, needs to be appropriate to the size of the project. On a project of 50 days it would be excessive to spend 20 days on project management. However, because a project is small that does not mean it requires no control and therefore has no need to follow a method.

The processes that need to be tailored relate to the following:

- planning and estimating;
- monitoring and control;
- issue management;
- change control;
- risk management;
- project assurance;
- project organization;
- business change management.

All of these will be described in greater detail in later sections of this book but a brief overview will be useful at this point.

### 1.6.1   Planning and estimating

Planning provides the basis for establishing control over a project. Planning must take account of all activities within a project. Good planning increases confidence within the project team. Without a plan there is no means of knowing whether dates can be achieved nor whether the project is adequately resourced.

The principal aim in planning is to detail all the activities, dependencies between activities and the resources required for the project to achieve its objectives. The plan shows when activities are to be performed and facilitates estimating of the staff effort needed and the most likely completion dates. An **outline** plan for the whole project will be produced initially but then a **detailed** plan for each stage will be created nearer the time that stage is to start. Chapter 2 looks at planning in more detail.

### 1.6.2   Monitoring and control

The project plans form the basis for monitoring progress against the expected achievements. Tracking and control aims to ensure that the project meets its commitments in terms of deliverables, quality, time and cost by ascertaining the current state of the project and identifying any need to re-plan. Chapter 3 examines project control in more detail.

### 1.6.3   Issue management

During the course of a project, problems will be identified which are considered likely to affect the success of a project. These problems or project issues may or may not be within the control of the project manager, but do not include authorized changes to the project requirements (see Section 1.6.4). The project manager should ensure that a system for recording these issues, monitoring their status and initiating any actions necessary is in place.

### 1.6.4  Change control

Any project will be subject to change during its lifetime. A change may result from a modification to requirements or as a result of errors found in testing. Any requests for change should be made through a formal change management process. Failure to incorporate necessary changes might reduce the benefit obtained from doing the project. However, accepting changes in an uncontrolled manner can cause problems with the cost, time scales and overall objectives of the project.

Change control ensures that any changes made are implemented with an understanding of the potential impact on the project and its business case. Chapter 4 examines change control and configuration management.

### 1.6.5  Risk management

All projects are subject to risk. If these risks are not managed then they can have a detrimental effect. A suitable risk management process assesses and manages project risks.

Risks are different from issues. A risk is an unplanned occurrence which could happen, but has not yet done so. An issue is an unplanned occurrence which has already happened and which requires the project manager to request or initiate action not previously planned.

Risk management identifies and quantifies risks before they happen, and plans and implements actions to eliminate risks or reduce their probability or impact. Risk management ensures that projects are only undertaken with a full understanding of the potential implications of the risks involved. Chapter 7 deals with risk in more detail.

### 1.6.6  Project assurance

Project monitoring and control involves monitoring various aspects of work, including progress, cost, changes, issues, risk and quality. When pressure mounts on a project to meet its deadline, it is tempting to ignore some of the checks and balances imposed by project control and to focus exclusively on the work to be done. This lack of control can be dangerous and can lead to project failure. Project assurance is a set of procedures which ensures correct project control is maintained. This may require auditing by staff outside the project team.

### 1.6.7  Project organization

A key factor in any project is an effective project organization where the roles and responsibilities of all participants are clearly defined and understood. Chapter 8 explicitly addresses this topic.

### 1.6.8  Business change management

Winning the support of stakeholders for a project is important for project success. Unless time has been spent communicating with stakeholders and making sure that users know exactly what to expect of the new system, the project could meet its formal requirements but still be seen as a failure by its users.

## 1.7 Development process models

Organizations involved in IT development need a well-defined, repeatable and predictable system development life cycle. Projects to create different types of product need different development life cycles. Whatever the life cycle, it will be a set of activities which, after completion, results in one or more products that are delivered to a customer. Each activity in the process will have a defined input and output.

Effective development methods should have certain characteristics. They should be made up of an overall set of techniques and activities from which team members working on a new project in that technical area can select the most appropriate set. The method should never require a task that does not produce something useful to the project.

The conventional system development life cycle for IT projects was described in Section 1.4. This has certain general characteristics that could apply to a number of different life cycles. It is assumed that each activity is normally done in a strict sequence, although there is some scope for reworking stages once they have been completed. This general structure is called a process model. The conventional system development life cycle conforms to a process model which is variously named the waterfall, one-shot or once-through model. The other process models that we will look at are the incremental and iterative process models.

### 1.7.1 Waterfall model

The waterfall model (see Figure 1.1) is the basic phased model of a development cycle. The model gets its name from the way each phase cascades into the next. There are variations but they are all based upon a systematic transition from one phase to the next until the project is complete. Phases should produce a sequence of deliverables such as the requirements statement, design documents and software structures, where the output from each phase is an input to the next.

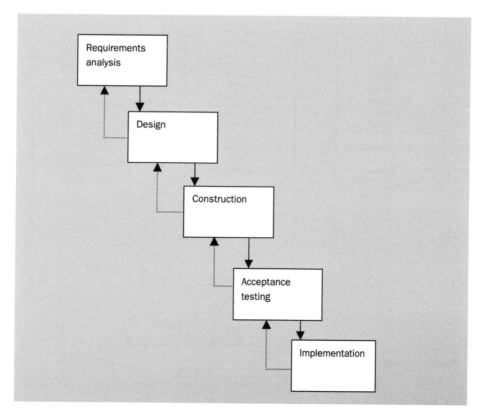

Fig. 1.1
The waterfall model

This approach provides for feedback loops which are activated when there is a need to revisit an earlier stage to redesign, recode and so forth. It is possible to return to any previous phase, although this could well require extensive replanning. The ideal is for quality control activities to be associated with each phase, meaning that once the deliverables of a stage have been signed off they should not need to be reworked.

This model is probably best used on projects where requirements have been clearly defined and agreed. Unfortunately, most projects do not have clear requirements at the beginning. As the model relies upon having each phase completed and signed off, it can become bureaucratic and time-consuming. It works best where there are few changes to requirements during the development cycle.

Other possible drawbacks include the amount of project documentation which can be created. The presence of a distinct testing phase at the end of the project might mean major defects are not found until late in the project when they are difficult to repair. It is also easy to misjudge progress: because the requirements have been signed off, it does not necessarily mean the requirements have been clearly understood.

### 1.7.2    Incremental model

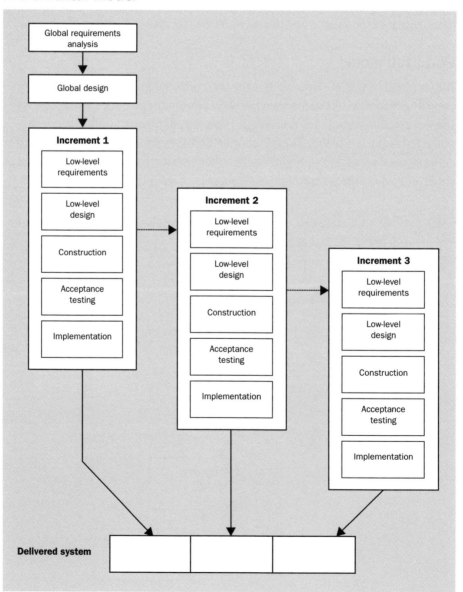

Fig. 1.2
An incremental model

Although the incremental model (see Figure 1.2) is similar to the waterfall model, it involves the development and delivery of functionality in fragments or **increments**. Typically, global requirements definition and design phases are completed and documented. Then the product is developed in increments. After each increment is designed, developed and tested, it is system tested and then becomes operational.

This approach works best when the requirements are relatively well-known. It can work well with larger projects, as these are effectively reduced to a series of mini-projects, each delivering an increment.

The incremental model is often used in conjunction with **time-boxes**. The deadline for completion of the increment is fixed and the features to be delivered by the increment are ranked according to importance. The least important features may be dropped to ensure that the deadline is met. The dropped features can be implemented in a subsequent increment if they are still required.

### 1.7.3 Iterative model

This model (see Figure 1.3) is suited to situations where the requirements are not clearly understood and where there is a need to begin development quickly to

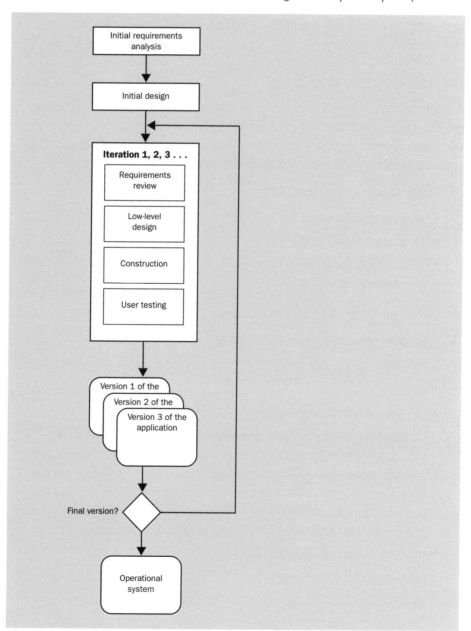

**Fig. 1.3**
An iterative model

create a version of the product which will demonstrate its look and feel. Early versions, or **prototypes**, of the system are created to help the customer identify and refine requirements and design features.

The iterative model can be used for some or all of the mini-projects where an incremental approach has been adopted.

Disadvantages associated with this model include not knowing when to stop iterating. The iterative approach is potentially difficult to monitor and control.

## 1.8   The project plan

During the project set-up phase (see Section 1.4.3), a project plan is produced that consists of several different types of document including activity networks and Gantt charts (see Chapter 2).

The project plan is a set of documents that co-ordinates all the various project processes. It brings together all the planning documents used to manage and control the project. It is not cast in stone and will be amended as necessary during the lifetime of the project. The plan defines the project scope, schedule and cost as well as the supporting processes related to risk, procurement, human resources, communication and quality.

### 1.8.1   Project initiation document

A good starting point is the preparation of a project initiation document (PID) or project management plan. Different project management methods give this document different names, but in essence it serves as an agreement between the sponsors and the developers of the project. It has the following key elements.

- An introduction, which describes:
  - the project background;
  - the document's purpose;
  - the business justification for the project, including a brief summary of costs and benefits (see Section 1.9).

- Project goals, objectives and deliverables.

- A project organization chart which names:
  - the project sponsor(s);
  - the project manager;
  - any sub-project or stage managers.

    The organization chart should detail the reporting relationships of all members of the project team and identify any steering committee (or project board or project management board) established for the project. It should show the reporting relationship between the steering committee and the project team. It should also specify the levels up to which named individuals, groups (such as the steering committee) or roles can authorize:
  - the commitment of resources;
  - the sign-off for documentation;
  - changes to goals, objectives, deliverables cost or time-scale.

    Chapter 8 discusses many of the issues involved in project organization.

- A project structure section that describes how the project will be broken down into manageable portions of work which will be administered as stages.

- A list of project milestones, i.e. significant events in the project for which dates need to be clearly specified. Milestones are used to measure the progress of a project and can be the start or completion of a major project phase. Milestones are events that in themselves consume no resources but a great deal of attention is paid to them by management at a senior level. Additional resources may be used in reviewing the state of the project before or after the milestone has been reached.

- Project success and completion criteria.

- A management control section that covers:

  - the frequency, timing, recipients and format of progress reports;
  - how the plan will be produced and maintained;
  - what information will be monitored and recorded;
  - how the information will be recorded;
  - how packages of work will be signed off and reviews conducted;
  - the people responsible for recording and assessing the impact of any changes;
  - the people responsible for authorizing different levels of change to goals, objectives, deliverables, cost or completion date.

- A risks and assumptions section that should identify any high-level risks to the project and propose specific actions to reduce or eliminate each risk. It is useful to include in this section a list of assumptions made in producing the report.

- A communication plan that provides an overview of how the project will communicate with the wider business organization, particularly with regard to changes needed in the business in order to make the implemented IT application effective.

- A report sign-off section.

  The document should be signed off by staff able to represent all areas and functions committing resources to the project and those who will be affected by the project. In so doing they implicitly accept the assumptions listed. Final sign-off should be obtained from the project sponsor. The project manager should not initiate any work which has not been explicitly or implicitly authorized in the project initiation document and been signed off by the project sponsor.

## 1.8.2 Schedule planning

Scheduling is probably the key project planning technique and must take place prior to the start of work. The level of resources required and the consequent costs incurred will depend on the schedule established for the work. Effective scheduling requires:

- a definition of requirements that is agreed and unambiguous;

- a careful breakdown of work;

- the creation of a coherent and internally consistent schedule which shows when activities will start and end and the resources that they will need;

- careful monitoring of progress against the schedule.

It is the responsibility of the project manager to make sure these requirements are fulfilled. Schedules will generally be produced at several levels of detail:

- project level, at which the project steering committee or board reviews plans and progress and takes decisions;

- phase/stage level, at which project or stage managers break down the main stages of the project into activities which are then allocated to teams;

- activity level, at which team leaders allocate work to team members so that activities allocated to the team can be completed.

As will be seen in Chapter 2, an important planning document for showing the schedule is the Gantt chart.

### 1.8.3   Cost planning

Once the activities that are to take place have been identified, the costs they incur can be assessed. Effective cost planning helps:

- decision-making and control by allowing the costs of possible actions to be assessed;

- prompt and up-to-date collection of progress information;

- integration of expenditure projections into the project plan.

It is necessary to estimate quantities and costs, set budgets and eventually control expenditure. Cost planning should be done at the same levels as schedule planning, that is, at project, phase/stage and activity level (see Section 1.8.2).

Controlling the costs of staffing is not easy. For example, time must be allowed for the acquisition of staff, if they are being recruited from outside the organization or if they need to be released to the project when their existing work commitments have been completed. If a delay is imposed on the start of an activity after the people who were to carry out the work have joined the team, the costs of those staff may still be charged to the project even if they are not yet working.

Project costs may be plotted on a cumulative resource chart (see Section 3.7.2).

### 1.8.4   Resource planning

The project plan needs to account for various types of resources including people, equipment and facilities. In many organizations, management must obtain resources from other parts of the company to work on a project. Functional managers who are responsible for various specialist departments will often be key resource providers. Through them resources like office space, computer equipment and specialist expertise can be obtained. If you need to obtain goods or services from outside the organization, the person responsible for contract management will be important.

Ultimately the success of a project depends upon the individuals involved and thus it is important to recruit the best people available. Unfortunately it is often the case that project managers have no choice but to use the staff allocated to them.

A resource plan will help the project manager identify the skills needed for the project and when they are needed. For example in many cases a project will not need a testing expert throughout its life but only at certain times.

There are several techniques that a project manager can use in resource planning. One example is the responsibility assignment matrix (RAM), a simple matrix showing individuals associated with the project on one axis and the activities for which they are responsible on the other. It is a useful communication tool for the project manager. By plotting tasks against staff on a matrix and labelling each person's assignments, it provides a quick and easy view of who is responsible for each task.

## 1.8.5  Communication planning

Stakeholders, including the project team, have differing needs for information about the project. These needs and the means by which they may be satisfied are recorded in a communication plan. The communication plan includes, among other things:

- the flow of communication during the project;
- how various communication tools will be used;
- what meetings will be held with what attendees and at what times.

## 1.8.6  Quality planning

In order to develop a system which complies in all respects with the users' business objectives and with the performance requirements documented during analysis, it is important to set up a carefully considered quality plan.

Quality criteria can be applied both to project deliverables and to the processes by which the deliverables are created. It is possible to check that the end products have the required qualities and also that the processes that created them were the correct ones. Because the importance of quality often gets lost under deadline pressures and budget cuts, the project manager should emphasize it throughout the entire project life cycle, from planning to completion.

The customer must be the final judge of the quality of the product, and must therefore be involved in quality evaluation. Chapter 5 further explores the role and content of the quality plan.

## 1.9  The business case

This is a key document for any project. As noted in Section 1.4.1 the initiation of a project may be triggered in a variety of ways. Once a proposal has been made, senior management will typically use a combination of qualitative and quantitative criteria to decide whether to go ahead with a proposal, as no single method gives enough information.

Qualitative criteria might include **organizational fit**: for instance, does the project fit with what the organization wants to do as part of its strategic mission? The **risk** associated with a particular project can also be assessed qualitatively at this point (see Chapter 7).

The main reason for going ahead with a project, however, tends to be a quantitative one of **financial justification**. Among the financial criteria that can be used for evaluation, two of the most common are **net present value** and **payback period**. In addition to these criteria the following may also be considered.

● Is there a legislative need that will be satisfied by this project?

● What impact will the project have upon the company's future capabilities?

● Is the project needed for growth?

● Is the project in alignment with long term strategies?

The contents of the document itself will vary but may include a description of the project, its objectives and scope, a cost benefit analysis, a risk analysis, a conceptual solution, resource requirements and success criteria. Some organizations include alternative options in the business case in order to compare the recommended solution against other approaches. The business case needs to be carefully reviewed by project sponsors.

## 1.9.1    Net present value

Note that for the purposes of the ISEB foundation certificate, you do not require knowledge of the method of calculating the net present value of a proposed project.

The financial business case is based on the calculation of whether the value of the benefits produced by the project exceeds the money spent on developing and installing the system and eventual costs of operation. The value of costs and income, however, is influenced by when they are incurred or received. The net present value technique looks at the benefits of having money sooner rather than later. Because money in hand can be invested to make more money, the earlier you bring money into the organization, the more it is worth.

Present value calculations translate future costs and benefits to a present day value using the formula

$$\text{Present Value} = \text{Future Value} \times (1/(1+i)^n)$$

where i is the interest rate (or discount rate) and n is the number of time periods (usually in years) from today. The discount rate is the amount of return that one could expect to receive elsewhere for an investment of comparable risk.

Say we were to receive £100 in one year's time and the interest rate was 10% (for ease of calculation!). The present value of that £100 would be

$$£100(1/(1.10)^1) \text{ i.e. } £90.90.$$

In other words, if I put £90.90 into an account at 10% interest then in a year I will have earned £9.09 which would give me a fraction under £100 in all.

The net present value of a proposed project is the present value of benefits (that is, money coming in) less the present value of investments (that is, money going out). The net present value technique allows an organization to estimate the value of money earned several years into the future and can be used to compare the expected results of investing in different projects with one another and with leaving money in a bank.

Here is a more detailed example, which uses a discount rate of 10%. Remember that for the ISEB foundation certificate you do not need to remember the details of the calculations.

A small organization decides to outsource its catering to outside caterers and invites companies to tender. Whichever caterer wins the contract will pay the host organization an agreed amount of money from the revenues that it will receive from charges for meals.

**Table 1.1  Net present value calculation**

| Simon's company | | | Christopher's company | | |
|---|---|---|---|---|---|
| Year | Cash flow | PV | Year | Cash flow | PV |
| 0 | -425 | -425 | 0 | 0 | 0 |
| 1 | 400 | 364 | 1 | 300 | 273 |
| 2 | 450 | 372 | 2 | 300 | 248 |
| 3 | 500 | 376 | 3 | 300 | 225 |
| Totals | 925 | 686 | | 900 | 746 |

The table above represents bids from the two contractors. It can be seen that Simon's company wants some money initially to help set up the operation. If you calculate the total cash flows for both proposals, then despite the initial reverse payment Simon's seems slightly better at £925 rather than £900. However when the figures are converted to reflect net present value, Christopher's is revealed as being much better at £746 as opposed to £686.

## 1.9.2  Payback period

Net present value calculations draw attention to the fact that money received early on in the delivered system's life cycle is more valuable than that received later. It is also the case that the forecasts of income become progressively less accurate as we look further into the future. Interest rates could change, or the economy could go into recession and undo our calculations. Thus how quickly you begin to make money is critical. This is what payback period calculations measure. From an organizational point of view, the shorter the payback period the better. Indeed, many organizations will be looking for a payback period of one year or less.

In the Canal Dreams booking system implementation project, say that the expenditure to set up the system is £10,000. This is allocated to year 0, which really means 'the period up to the system going live'. In the table below, the income is made up of the value of the additional bookings that the new system allows and the savings in booking staff. It can be seen the accumulated cash flow becomes positive somewhere near the middle of year 3, so that the payback period is about 2.5 years.

**Table 1.2  Payback period calculation**

| Year | Expenditure | Income | Annual cash flow | Accumulated cash flow |
|---|---|---|---|---|
| 0 | 10000 | 0 | -10000 | -10000 |
| 1 | 1000 | 4000 | 3000 | -7000 |
| 2 | 1000 | 5000 | 4000 | -3000 |
| 3 | 1000 | 7000 | 6000 | 3000 |
| 4 | 1000 | 7000 | 6000 | 9000 |

## 1.10    Implementation strategies

At the other end of the project, there will come a time when it is necessary to convert from the old method of working to the new one brought about by the implementation of the new system. The project team, in consultation with the users, will recommend to the project board or steering committee the most suitable changeover method for the project, which may include installing IT equipment, software applications or both. The following options may be considered.

### 1.10.1    Direct changeover

In this case, the old system is discarded and is immediately replaced by the new one. It can be considered a risky approach but is relatively inexpensive if thorough testing has been done. The more complex and important the new system, the riskier this approach will be, especially if there is no possibility of falling back on the old system in the case of failure.

### 1.10.2    Parallel-running

This involves running the old and new systems together for a period of time using similar inputs and comparing the related outputs – so it serves as a continuation of the testing process. It is a safe, low-risk approach but can be expensive, particularly in terms of duplicated labour costs. Problems can also occur in determining how long to run the two systems in parallel.

### 1.10.3    Phased take-on

The phased approach breaks the system into components that will be introduced in sequence. It helps minimize risks but can delay the implementation of the entire integrated system. However it does present the opportunity to allow users to learn one system component at a time. It also fits neatly with incremental delivery.

### 1.10.4    Pilot changeover

Like the phased approach this is a risk-reducing approach. With pilot changeover the entire new system is introduced to one business unit or location at a time. It can only be used if the business unit or location can use the entire system independently. Problems can be addressed and fixed before the system is introduced company-wide but company-wide deployment of the entire system is consequently delayed.

**What would be the best implementation strategy for Canal Dreams?**          *Activity 1.4*

## 1.11    Post-implementation review

The **post-implementation review (PIR)**, or **project evaluation review**, is usually scheduled to take place some 6 to 12 months after the sign-off of the project. Its objective is to review the implemented system in terms of its contribution to business objectives, its usability, operating costs and reliability. It considers the following:

● whether the business and system requirements have been met;

- cost and benefit performance;
- operational performance;
- controls, auditability, security and contingency;
- ease of use.

The output from this process is a **post-implementation review report**. The review should be led by someone who is independent of the project and should solicit feedback from users, operations and the support team. The review should address the operational system and not the development project. In view of the effort involved, the users might not be committed to this review. If this is the case, it is essential to explain to the users the benefits of the process, for example that changes that could improve the system may be identified as a result.

# Sample questions

1. Which of the following is NOT a characteristic of a project?
   (a) Ongoing nature
   (b) Uniqueness
   (c) Clear objectives
   (d) Integration of interrelated tasks and resources

2. Which of the following is NOT managed by the project manager?
   (a) Time, cost and scope
   (b) The project team
   (c) The project sponsor
   (d) Expectations of the stakeholders

3. Which of the following tools indicates who is responsible for what?
   (a) A responsibility assignment matrix
   (b) A resource levelling chart
   (c) An activity network chart
   (d) A resource histogram

4. With which one of the following does the calculation of the payback period provide the organization?
   (a) An assessment of how quickly an implemented system will produce a profit.
   (b) How much future income from an implemented project will be worth in present day terms.
   (c) The discount rate that would give a payback with a net present value of zero.
   (d) The overall income that the implemented system should produce, less any initial investment.

## Answers to sample questions

1. (a)   2. (c)   3. (a)   4. (a)

## Pointers for activities

### Activity 1.1

Among the activities that may be considered are the following.

(i)     Survey existing office requirements (for example, how many desks and chairs there are) and IT infrastructure, including servers and printers, used by the office.

(ii)     Survey new office space.

(iii)     Plan new office layout – who and what will go where?

(iv)     Schedule the sequence of moves – it might be that not all staff should be moved at once as this will allow for some continuity of service.

(v)     Select a removal company.

(vi)     Organize the setting up of the infrastructure.

(vii)     Pack up the old office.

(viii)     Transfer to the new office, including supervision of placement of furniture etc.

(ix)     Unpack in the new office.

(x)     Connect telephones, etc.

As project manager you will probably want to be consulted when decisions have to be made, possibly at points (iii), (iv) and (v) as money is involved here. Useful checkpoints would be just before (vii) to check everything is in order to go ahead and after (x) to check out any outstanding problems.

### Activity 1.2

There will be some interviews with fairly high-level managers, including the project sponsor (who may be the managing director, who also owns the company), who will be able to clarify the business requirements of the proposed system. Staff who currently carry out the clerical booking operation would need to be interviewed to see how the system currently works, what data has to be held and what problems there are with the current system. Operations at more than one boatyard would need to be studied, especially since they were originally different businesses, as there could well be local differences in procedures. Staff in areas adjacent to the booking operation may need to be approached, to document the interfaces between the booking operation and other parts of the overall Canal Dreams business, e.g. the central finance function and the maintenance staff who may need to schedule boat maintenance.

### Activity 1.3

Among the input screens that may be needed are:

- set up new boat details
- remove boat

- record provisional boat booking
- cancel boat booking
- confirm boat booking
- record payment
- set up prices

## Activity 1.4

Canal holidays is a seasonal business with very busy times of the year and times in the off-season when the business is dormant. This would seem to argue in favour of a direct changeover. A pilot changeover, where the new system was only used at one boatyard to start with, could also be considered.

# 02

# Project planning

## LEARNING OUTCOMES

When you have completed this chapter you should be able to demonstrate an understanding of the following:

▶ project deliverables and intermediate products;

▶ work and product breakdowns;

▶ product definitions (including the identification of 'derived from' and 'component of' relationships between products);

▶ relationship between products and activities in a project;

▶ checkpoints and milestones;

▶ lapsed time and effort required for activities;

▶ activity networks (using 'activity on node' notation);

▶ calculation of earliest and latest start and end dates of activities and the resulting float;

▶ identification and significance of critical paths;

▶ resource allocation, smoothing and levelling, including the use of resource histograms;

▶ work schedules and Gantt charts.

## 2.1 Introduction

In this chapter we describe the main steps in producing an initial plan for a project. You will recall from Chapter 1 that before the detailed planning of a project starts, the **business case** for the project should have been set out. This shows how the benefits of the proposed IT application are expected to outweigh the costs of developing and managing it. The overall **objectives** of the project, which define the successful outcomes of the project, have also been identified and have been agreed by the main participants in the project.

## 2.2 Approaches to planning

There are two approaches to identifying the components of a project: **product-based** and work- or **activity-based**.

### 2.2.1 Product-based planning

With the product-based approach, detailed planning usually begins with identifying the **project deliverables**, that is, the products that will be created by the project and delivered to the client. A product must be in some way tangible, but it can be any of a wide range of things. It could be a software component, a document, a piece of equipment or even a person (for example, a trained user). It could be a new version of some existing product, as with a modified version of a software component.

In the case of the Canal Dreams project, the deliverables may include:

- a central, suitably furnished office for bookings;
- IT equipment;
- a software application that can be used for bookings;
- a training manual;
- trained staff;
- office procedure manuals;
- a database primed with details of the boats that can be booked.

Once the deliverables have been defined, **intermediate products** can be identified. These are products which are created during the course of the project, but which may not actually be delivered to the client at the end of the project. In the case of the Canal Dreams scenario, intermediate products may include:

- software specifications;
- a database structure;
- acceptance test plans;
- acceptance test reports;
- progress reports.

Some products, such as the progress reports in the list above, may relate to the management or quality control of the project.

The intermediate products can be written up simply as a list of products, but sometimes they are portrayed in the form of a **product breakdown structure**.

Some of the stakeholders in the project may find there are some products with which they are unfamiliar. Some users, for example, may be unsure of what is meant by an 'acceptance test plan'. To remedy this, planning should include drawing up **product definitions.** For each product, the following should be documented:

- The **identity** of the product, for example 'acceptance test plan'.

- A **description** of the product, for example 'a plan of the test cases and the results that the users expect the application to produce'.

- The product or products that have to exist before this one can be created, that is, those it is **derived from**. For example, the acceptance test plan may be derived from the requirements specification, which states what the main transactions of the application are going to be.

- The **components** that make up the product. In the case of an acceptance test plan, these may be the main sections in the document.

- The **format** of the product, for example that it is a word-processed document or a spreadsheet or a piece of software.

● **Quality criteria** that explain the processes by which the product will be judged satisfactory, for example, by being reviewed against the requirements specification.

## 2.2.2 Work and product breakdown structures

An alternative method of planning is the activity-based approach, which starts by identifying the work activities or tasks that are needed in a work breakdown structure. In this case, the products related to setting up the Canal Dreams central booking office (see Figure 2.1) would be replaced by activities such as:

● allocate room;

● install office furniture;

● install IT equipment;

● train staff;

● draft office procedures.

As nearly all activities will generate a product – or else, why do them? – and all products will need to have some activities that give birth to them, there may not really be much difference between the two approaches in practice.

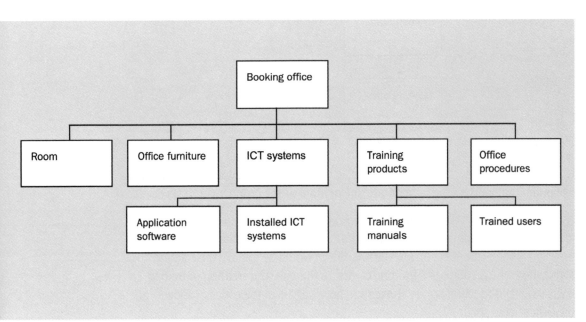

Fig. 2.1
A product breakdown structure diagram

**Which products are created by each of the following activities?**

(a) testing

(b) training

(c) network installation

(d) a project progress meeting

*Activity 2.1*

## 2.3  Product flow diagram

If you have adopted a product-driven approach, it is possible to draw up a product flow diagram (PFD) showing the order in which products have to be created. This should be relatively easy to draft if you have already produced product descriptions that specify from which other products each product is derived. Figure 2.2 gives an example fragment of a product flow diagram.

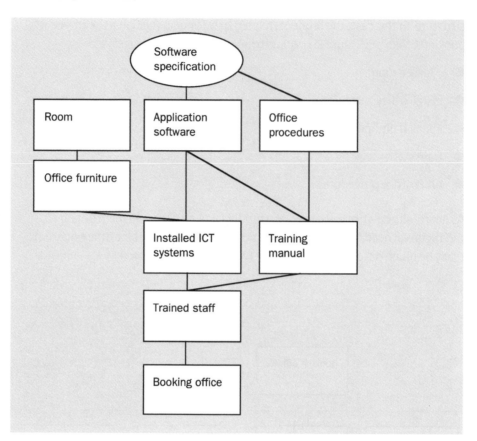

**Fig. 2.2**
A product flow diagram

Note the oval with 'software specification' in it. This refers to a product that already exists and that will be used to generate one or more of the products in the PFD.

There is no one correct PFD – its structure depends on the decisions made. For instance, in Figure 2.2 the assumption is that training will take place in the office once it has been equipped. An alternative option would be to have the training done somewhere else, perhaps in a training centre. In this case the creation of trained users would not depend on the application being set up in the office or, indeed, on the office being set up.

## 2.4  Activity planning

Whether or not a product flow diagram has been drawn up or whether the planner has simply drawn up a list of activities, the next step is to draw up an **activity network**. This shows the activities needed and the order in which they are to be carried out.

### 2.4.1  Activity network diagram

There are two sets of conventions for drawing up activity networks: 'activity on node' and 'activity on arrow'. Figure 2.3 shows an example of an activity on arrow diagram.

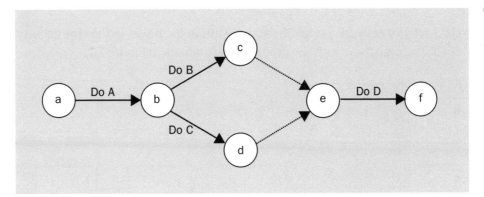

Fig. 2.3
Activity on arrow
network

As the name implies, the arrows in an 'activity on arrow' diagram represent activities, while the circles that link the arrows (that is, the **nodes**) represent the ends of some activities and the starts of others. The arrows with broken lines indicate 'dummy activities' which simply show a dependency between two of the event nodes, for example c, the end of 'do B', and e, the start of 'do D'.

In these notes we use a different set of conventions, **activity on node**, which is used by most modern project planning tools, for example, Microsoft Project.

Figure 2.4 shows the same activities as Figure 2.3, but using activity on node notation. Here the boxes (which are the 'nodes' in this case) represent activities while the lines between the boxes show where the start of one activity depends on the completion of some other activity. Note that at this stage the constraints may be technical or external. A technical constraint normally means that a product has to be created by one activity so that another can use it. An example of an external constraint is a system that can only be tested out of office hours, so it has been agreed contractually that testing will take place at weekends only.

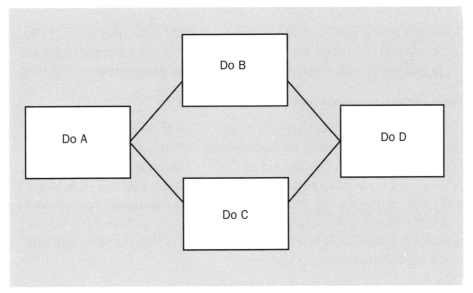

Fig. 2.4
Activity on node
network

What you do not take into account at this point are **resource constraints,** for example that a person will not be able to start on one task because he or she will not have finished another. These considerations are deferred because a decision may be taken, when staff are being allocated, to employ more staff.

In this activity network, match the activities with the boxes so that the activity    *Activity 2.2*
network is compatible with the product flow diagram in Figure 2.2.

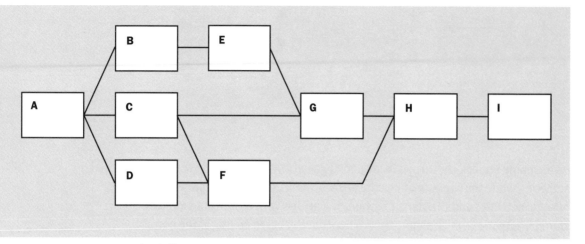

...... install computer systems          ...... start

...... train staff                       ...... finish

...... write/test software               ...... allocate room

...... draft office procedures           ...... install office furniture

...... write training manual

In Activity 2.2 we added two 'activities' called 'start' and 'finish'. These are important points of time (or 'events') in the life of the project, but they will not actually take up any time. If, for example, the finish of the project was marked by a celebration that took up several hours, then the 'event' would become an activity in its own right. We call these important events **milestones**. Milestones can also be located in the middle of a project, for example at the end of one important phase and the start of the next. Always remember, though, that milestones do not take up time. Sometimes an important point in a project may be marked by a meeting to check that everything planned has been completed successfully before the next part of the project starts. This checkpoint would be an activity in its own right.

### 2.4.2   Estimating elapsed time

Having identified the activities and the order in which they have to be worked on, we now need to estimate how long we think each activity is likely to take. Note we are concerned here about **elapsed times.** This is the time from the start of an activity to the finish. This is not the same as the **effort** spent on an activity. Effort could be more than elapsed time, for example where we have three people working on a job for two days, the elapsed time would be two days but the effort would be six staff-days. Effort could be less than elapsed time, for example where someone works only afternoons.

Let us assume that we can allocate estimated durations to the activities in the activity network of Figure 2.4 (see Figure 2.5).

We want to calculate the earliest day upon which each activity can start. Rather than worry about taking account of weekends and bank holidays at this point, we simply allocate each day a number, starting with day 0. (Technically, day 0 means 'the end of day 0' which means the start of day 1, as explained below.)

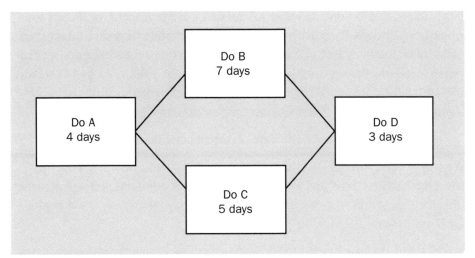

Fig. 2.5
A network activity
diagram with activity
durations

The **earliest start date** for 'Do A' is day 0 by definition because it is the first activity in the network. The earliest the activity can finish is on day 0 plus the duration of the activity, that is day 4.

earliest finish date = (earliest start date + activity duration)

The earliest start dates for the two activities 'Do B' and Do C' are governed by the earliest finish date of the preceding activity 'Do A'. In fact we can say that in this case the earliest start dates for 'Do B' and 'Do C' are the same as the earliest finish date for 'Do A'.

You may wonder why it is not the **following** day. Well, the convention is that when we say that 'Do A' finishes on day 4 we mean at the **end** of day 4. When we say that 'Do B' and 'Do C' start on day 4 we really mean at the **end** of day 4, which of course really means the start of day 5. It is best just to accept this as the convention. It saves problems where activities do not take whole numbers of days, for example 5.5 days.

We can now work out the earliest finish days of 'Do B' and 'Do C' as day 11 and day 9, respectively. What about the earliest start date for 'Do D'? We have two preceding earliest finish dates so we take the one which is later, that is day 11.

earliest start date = the latest of the earliest finish dates of the preceding activities upon which the current activity is dependent

We end up with the day numbers shown in Figure 2.6, where ES means the earliest start date and EF means the earliest finish date.

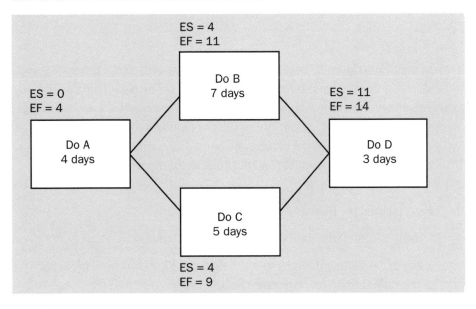

Fig. 2.6
Earliest start and
finish days

It may be possible for some activities to start or finish late without the project as a whole being delayed. To see where this is the case the **latest finish** and **latest start** dates for each activity are calculated. We will assume that we want the project as a whole to take the shortest time possible, that is, to finish on day 14. Day 14 becomes the latest finish date for the activity 'Do D'. The latest start day is calculated by subtracting the duration from the latest finish, that is, 14 – 3 = Day 11.

> latest start date = latest finish date of current activity – duration

We now work backwards. The latest start day for the activity 'Do D' becomes the latest finish day for 'Do B' and 'Do C'. By subtracting the durations for these activities from their latest finish days we get their latest start days, that is, 11 – 7 = day 4 for 'Do B' and 11 – 5 = day 6 for 'Do C'.

In the case of 'Do A' we have to decide whether to base the latest finish on the latest start date of 'Do B' or 'Do C'. The earlier of the two is taken as the latest finish time for 'Do A', that is day 4, which comes from 'Do B'.

> latest finish date = the earliest of the latest start dates of the activities that are dependent on the current activity

We now have the situation shown in Figure 2.7, where LS means latest start and LF means latest finish.

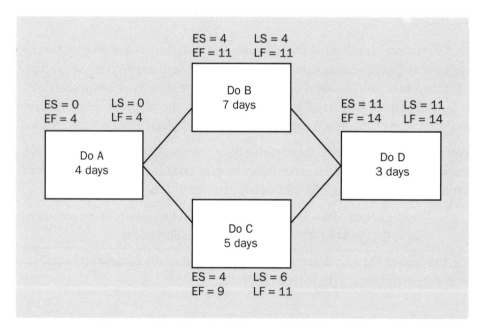

**Fig. 2.7**
Latest start and finish days

It can be seen from Figure 2.7 that for all the activities except 'Do C', the earliest and latest **start** days are the same and the earliest and latest **finish** days are also the same. This means that if these activities are late, then the project as a whole will be delayed. In the case of 'Do C', if you look at the day numbers you can see there is a two-day difference between the earliest and latest day numbers. This means that 'Do C' could be one or two days late and the duration of the project as a whole would not be affected.

This leeway is called the **float** and can be defined as:

> float = latest finish date – earliest start date – duration

A quick way of calculating this is by subtracting the earliest start from the latest start (or the earliest finish from the latest finish).

In Figure 2.7, 'Do A', 'Do B' and 'Do D' all have zero float. They form a small chain of three activities from the beginning to the end of the activity network. This chain is the **critical path.** If any activity on this path is delayed then the whole project will be delayed.

The details for each activity can be displayed more clearly if the boxes on the activity diagram are divided up as shown in Figure 2.8.

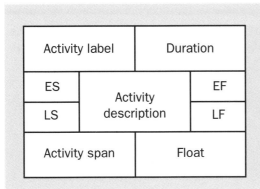

**Fig. 2.8**
Layout of an activity box

Thus for the activity 'Do C', the activity box in the activity network could be drawn up as in Figure 2.9.

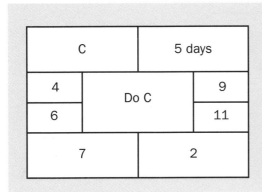

**Fig. 2.9**
Activity box for 'Do C'

The **activity span** is the total period during which the activity has to take place, and is defined as:

activity span = latest finish day – earliest start day

In this case it is 11 – 4, that is 7 days. Where an activity has float, there is a 'window of opportunity', reflected by the activity span (see Figure 2.10), within which the activity can start and be completed.

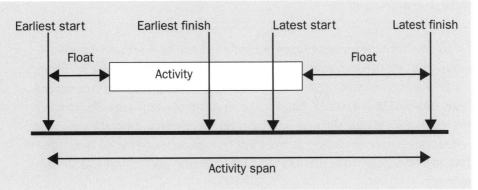

**Fig. 2.10**
Activity span

The activity has to take place within the period designated as the activity span. Because there is some float there is some freedom about when it can take place within that period. However, the start must be in the period between the earliest start and latest start. If it is not, it will not be completed within the activity span – unless its duration can be shortened in some way.

**Calculate the earliest and latest start and finish days and floats for each of the activities in the activity network below. Use the results to identify the critical path.**

*Activity 2.3*

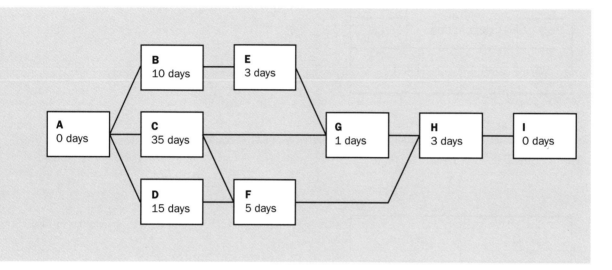

## 2.5  Resource allocation

So far we have taken no account of the availability of the resources needed to carry out any task. It is assumed that they will be available when they are needed. The resources that now need to be considered include raw materials, staffing and equipment. Usually with IT projects, the main concern is with staffing; however, sometimes equipment can also cause problems. For example, when conducting acceptance testing for a modified IT application, there is often a need for some tests of the whole operational system. This may need to be done on a public holiday or a weekend when no normal operational use is being made of the system. Here, we focus on staff resourcing.

For each activity, the **resource types** needed are identified. A resource type is a group of people of which any member could carry out a particular task. For example, if a software component needs to be written in Java, identifying Ali as the needed resource would be too precise; Jane may be equally proficient in Java. Identifying the resource simply as a software developer, on the other hand, may be too vague: Alfred is a software developer but, as a COBOL programmer, he has no knowledge of Java. Identifying the required resource as a Java programmer may be just about right.

Match the following resource types and activities.

*Activity 2.4*

| Activities | Resource types |
|---|---|
| allocate room | business analyst |
| draft office procedures | business solution developer |
| install computer systems | IT technician |
| install office furniture | premises manager |
| train staff | training consultant |
| write training manual | |
| write/test software | |

In order to illustrate the process of resource levelling and smoothing, we break down one of the activities in our Canal Dreams project, write/test software, into more detailed tasks (see Figure 2.11).

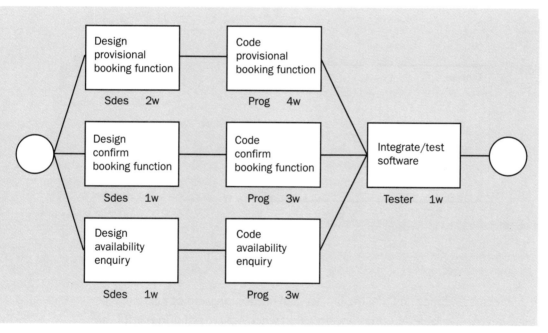

**Fig. 2.11**
Canal Dreams project: write/test software activity
SDes = system designer, Prog = programmer, w = week

Having allocated resource types to activities, we now go through the activity network and for each unit of calendar time (in this case each week) note the resources needed if the activity is to start at its earliest start date (see Table 2.1). We can also depict this information as a **resource histogram** (see Figure 2.12).

**Table 2.1  Quantity of each resource type needed each week**

| Week | 1 | 2 | 3 | 4 | 5 | 6 | 7 |
|---|---|---|---|---|---|---|---|
| System designers | 3 | 1 | 0 | 0 | 0 | 0 | 0 |
| Programmers | 0 | 2 | 3 | 3 | 1 | 1 | 0 |
| Testers | 0 | 0 | 0 | 0 | 0 | 0 | 1 |

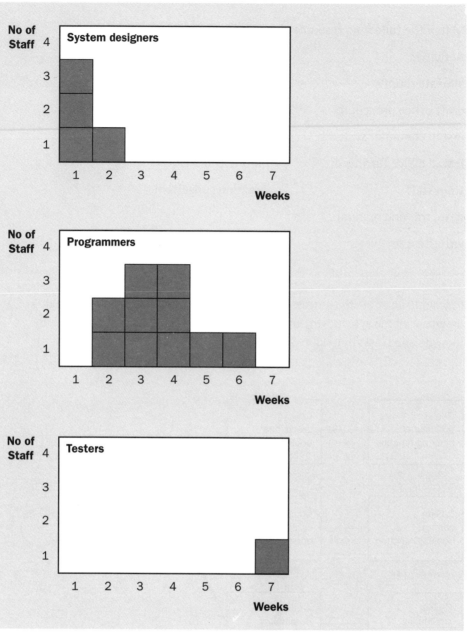

Fig. 2.12
A resource histogram for
each resource type

In some cases, we may find that the project plan needs more staff of a certain type than we have available. Another concern is that where we are acquiring staff from outside, we want to make sure that, as far as possible, we are able to provide a steady flow of work. We try to avoid having short periods of intense activity, where staff may need to be contracted in on expensive temporary contracts, alternating with periods when nothing much is happening and our permanent staff are under-employed. As well as the additional costs, the lack of familiarity of temporary staff with the project may mean that they are less productive at first.

Delaying certain activities may allow peaks and troughs to be 'smoothed', which will lead to more economical staff costs and a more productive work distribution. For example, in Figure 2.12, three system designers are needed in week 1 and only one in week 2. If we have only two system designers on the staff, then we could employ a third system designer, possibly on an expensive temporary contract, for the first week. If we calculate the float for each activity in the activity network (Figure 2.11), we can see that two of the activities needing system designers have two weeks' float. If we delay starting one of these activities until week 2, this part of the project as a whole will not be delayed but we will need only two system designers.

**Redraw Table 2.1 to take account of a delay of one week to the activity 'design availability enquiry'.**

*Activity 2.5*

When there are not enough staff of a particular resource type to carry out the activities due to take place at a certain time then there is a **resource clash**. Even if there are no resource clashes, a planner should try to arrange activities so that the usage of a resource type is as stable as possible.

Where there is a resource clash or the demand for a resource is very uneven, the following options may be considered:

- Use the float of an activity to delay the start of some work until the required staff member is available (as in the example above).

- Delay the start of an activity even though the float has been used up. This will delay the overall completion date of the project, but that may be preferable to the extra cost of employing more staff.

- Buy in additional staff to cover the staff deficiency. This will normally increase the cost of the project.

- Split an activity into sub-activities. For example, it may be possible to split the provisional booking function into two component sub-functions, each requiring a week of design and two weeks of programming. This could allow the demand for systems design to be spread more evenly.

**Redraw Table 2.1 to reflect the demand for the different types of resource if the provisional booking function is split into two equal-sized software components and a management decision is made to employ only one systems designer.**

*Activity 2.6*

If an activity has to be delayed until a member of staff is made available on the completion of some other activity, this should **not** be indicated by a dependency link between the two activities. Rather, the start date of the waiting activity should simply be amended. If other staff were to be released earlier than foreseen they could be used to expedite the waiting activity. If the waiting task had a dependency link to another task, it would imply a technical reason for waiting until the other was complete and would mask the opportunity to use other staff.

We are now in a position to put our plan into a form that will be easily understood by all those who are going to be carrying it out. The most common format is a Gantt chart (see Figure 2.13). The activities are listed down the left hand side and the calendar units (in this case, weeks) are shown along the top. In the body of the diagram there are block symbols for each activity, showing when the activity will be carried out. In the diagram, the free float related to each activity is indicated by the lighter blocks that extend the base period for an activity.

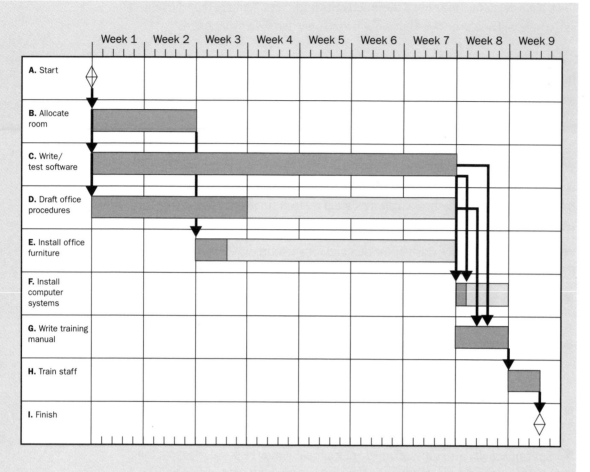

Fig. 2.13
Gantt chart

For a smaller project, an alternative layout is to list each member of staff down the left and to show what they are doing in each time period in the body of the diagram. This is not dissimilar to the holiday planning charts on the walls of many offices showing when staff will be away. A disadvantage of this format is that activities involving more than one team member have to be duplicated.

What does GANTT stand for in the name 'Gantt chart'?

*Activity 2.7*

## 2.6 Using software tools for planning

In this chapter it has been assumed that all the planning and the calculation of the consequences of particular planning decisions will be done by hand with no computer assistance. It will be a relief for most people to know that there are software packages that will carry out most of the calculations for you. Examples of these are Microsoft Project and Pertmaster.

In most cases, for each activity you will input the activity name, the duration of the activity, the activities upon which it is dependent and the resources that it will use.

Given this information, the software then produces activity networks, resource histograms, Gantt charts and other useful reports. Some packages suggest ways of resolving resource clashes, but users need to check that the results are what they really want. Where activity networks and Gantt charts are produced, the planner often has to tweak them to make them easy for others to understand. For example, a Gantt chart often spreads over several pages many of which are blank. The advantage of using a software tool is that it is easy to make changes to the plan and see what the consequences of the changes will be.

# Sample questions

Questions 1, 2 and 3 are about the diagram below. The number of days shown in each box is the duration of the activity.

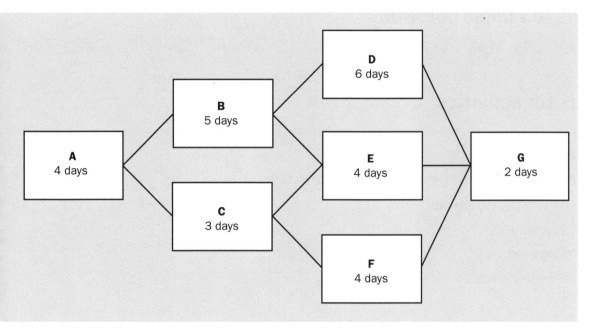

1.   Which of the following is the critical path?

(a)   A, B, D, G

(b)   A, B, E, G

(c)   A, C, E, G

(d)   A, C, F, G

2.   What is the float for activity F?

(a)   0 days

(b)   2 days

(c)   4 days

(d)   6 days

3.   Activities B and C have to be completed by the same person. What is the delay in the end time of the project?

(a)   3 days

(b)   5 days

(c)   4 days

(d)   1 day

**4.** Which of the following does not take account of the dependencies between activities?

(a) Gantt chart

(b) activity network

(c) work breakdown structure

(d) resource histogram

## Answers to sample questions

1. (a)    2. (c)    3. (d)    4. (c)

## Pointers for activities

### Activity 2.1

Among the products that may be created for each activity are:

(a) test results, error reports

(b) trained users

(c) a network

(d) meeting minutes, to-do lists, updated plans

### Activity 2.2

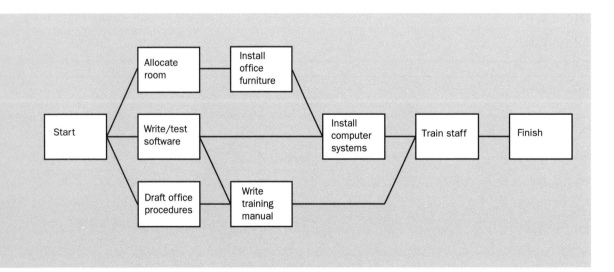

## Activity 2.3

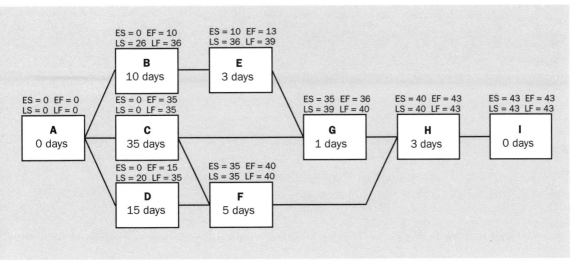

| Task | Depends on | Earliest start | Duration | Earliest finish | Latest finish | Latest start | Float |
|------|-----------|---------------|----------|----------------|---------------|-------------|-------|
| A | | 0 | 0 | 0 | 0 | 0 | 0 |
| B | A | 0 | 10 | 10 | 36 | 26 | 26 |
| C | A | 0 | 35 | 35 | 35 | 0 | 0 |
| D | A | 0 | 15 | 15 | 35 | 20 | 20 |
| E | B | 10 | 3 | 13 | 39 | 36 | 26 |
| F | C,D | 35 | 5 | 40 | 40 | 35 | 0 |
| G | C,E | 35 | 1 | 36 | 40 | 39 | 4 |
| H | F,G | 40 | 3 | 43 | 43 | 40 | 0 |
| I | H | 43 | 0 | 43 | 43 | 43 | 0 |

The critical path is A, C, F, H, I.

## Activity 2.4

| Activity | Resource type |
|----------|---------------|
| Allocate room | Premises manager |
| Draft office procedures | Business analyst |
| Install computer systems | IT technician |
| Install office furniture | Premises manager |
| Train staff | Training consultant |
| Write training manual | Training consultant |
| Write/test software | Business solution developer |

## Activity 2.5

| Week | 1 | 2 | 3 | 4 | 5 | 6 | 7 |
|------|---|---|---|---|---|---|---|
| System designers | 2 | 2 | 0 | 0 | 0 | 0 | 0 |
| Programmers | 0 | 1 | 3 | 3 | 2 | 1 | 0 |
| Testers | 0 | 0 | 0 | 0 | 0 | 0 | 1 |

## Activity 2.6

| Week | 1 | 2 | 3 | 4 | 5 | 6 | 7 |
|---|---|---|---|---|---|---|---|
| System designers | 1 | 1 | 1 | 1 | 0 | 0 | 0 |
| Programmers | 0 | 1 | 2 | 3 | 3 | 1 | 0 |
| Testers | 0 | 0 | 0 | 0 | 0 | 0 | 1 |

## Activity 2.7

GANTT does not stand for anything. Gantt charts are named after their inventor, Henry Gantt. Gantt should, therefore, not be written in capitals!

# CHAPTER THREE

# Monitoring and control

## LEARNING OUTCOMES

When you have completed this chapter you should be able to demonstrate an understanding of the following:

▶ the project control life cycle including planning, monitoring achievement, identifying variances and taking corrective action;

▶ the nature of and purposes for which project information is gathered;

▶ how to collect and present progress information;

▶ the reporting cycle;

▶ how to take corrective action.

## 3.1 Introduction

Chapter 1 described the typical stages of a project that implements an information system. The importance of controlling the project, that is ensuring that it conforms to the plan, was stressed. In Chapter 2, the way in which the plan for a particular project is created was explained. This chapter explores the means by which a project is monitored and controlled so that it broadly fulfils its plan. The mechanism for this is the project control life cycle.

## 3.2 The project control life cycle

The project control life cycle involves the following sequence of steps:

(a) Producing a plan for the project to follow;

(b) Monitoring progress against the plan;

(c) Comparing actual progress with the planned progress;

(d) Identifying variations from the plan;

(e) Applying corrective action if necessary.

Steps (b) to (e) are repeated to continue the **control cycle**.

Imagine a ship's voyage across the Channel from Dover to Calais as your project. The plan would probably involve following a certain route aiming to arrive in Calais at a certain time. As the voyage progressed, the navigator would check the ship's progress against the planned course. If there was a difference, he or she could then decide that a change of speed or an alteration of course was necessary – this would be **corrective action**. The process would, of course, continue until the ship arrived at its destination. Without this **control cycle**, the ship could continue on a fixed course and speed and would be very unlikely to arrive at the planned destination or at the expected arrival time.

In an IT project, monitoring progress is more difficult than in the ship example. The first question, which we tackle in the next section, is how to identify things that should be monitored. We usually know what the final objective of the project is but how do we know how well we are progressing towards that objective?

## 3.3   Monitoring progress

### 3.3.1   What should we monitor?

The most obvious thing to monitor is the **progress** in creating **deliverables** and other, intermediate, project products and in meeting **milestones** or **deadlines**. Difficulties however arise when you want to monitor progress and things are only partially complete. The stock answer is to break the products and deliverables into smaller components that can be assessed as complete at shorter and more frequent intervals of time. This may be quite difficult in IT projects. One established way is to break software down into segments or components. A common alternative is to try to assess the percentage completion of an activity or deliverable. This again is problematic. If someone is building a wall, it is easy to see when it is half finished although even then the finishing touches may take a little extra time. In software development it is quite common to take a long time to iron out the final snags in an 'almost finished' product. Software is also less obviously visible than a wall.

In Chapter 6, we describe **size** or **effort drivers**. These allow us to measure the size of the job to be done. In the case of building the wall, the number of bricks would be an obvious size driver: the bigger the wall, the more bricks it will need. The size/effort driver can be used to monitor progress. For example, if we know that the bricklayer will need to lay 200 bricks to build the wall but only 50 have been laid so far, then we can assume that the job is about 25% complete.

There are 20 boatyards owned by Canal Dreams, and staff at each boatyard will need to be trained in accessing the new booking system when it is implemented. It has been estimated that the trainer, on average, will need a day to travel to a boatyard, set up the training environment and conduct the training. 20 days (or four working weeks) are allocated for the training. However, at the end of the first week only three boatyards have in fact been visited.

*Activity 3.1*

(a) How long is it likely that the training will now take?

(b) What difference to the figure you have produced in (a) may the following circumstances make?

   (i)    The trainer was two days late starting because printed materials needed for the training sessions had not yet arrived from the printers.

   (ii)   The trainer started with the boatyards furthest afield and needed an extra day to travel to the area and travel back.

Another, less obvious, thing that needs to be monitored is the **use of resources**, which in IT projects means mostly 'human resources' or staff time. In addition, **financial expenditure** should be carefully monitored. Surprisingly, however, financial expenditure on human resources is not always strictly monitored in IT projects. This is particularly the case in situations where the project personnel are permanent employees in an IT department and are therefore viewed as overheads.

In addition to delivery time and cost, the **nature** of **deliverables** needs to be monitored, both in terms of the size or **scope** of the deliverables and their **quality**. One danger is that the amount of functionality to be delivered increases because new requirements are discovered. If these additions to the work are not monitored and controlled, clearly costs and delivery time are going to be affected. A further danger is that a task may appear to be completed, but in fact the poor quality of the resulting product means that the activity has to be re-opened in order to correct defects.

These three factors, of time, cost and deliverables, need to be balanced. For example, it may be possible to accelerate the progress of a project that is in danger of not meeting its deadlines by bringing in more staff, but this would increase the project cost. On the other hand, it may be possible to meet the deadline within the budgeted cost by reducing the number of features in the application that is to be delivered.

### 3.3.2 How should we monitor?

There are formal and informal methods and it is important to use both types. **Formal monitoring methods** may include the use of written reports, email and progress meetings. The frequency, format and content of these kinds of communication should be laid down formally at the start of a project in a project initiation document or its equivalent.

The advantage of **formal monitoring methods** is that routines can be established so that people are periodically forced to focus on progress and to commit themselves in writing. The disadvantage is that time spent preparing reports can be seen as an unproductive overhead. Staff need to be convinced of the value of reporting in order to gain commitment. For example, the use of **timesheets** can be effective in establishing the staff effort expended on distinct aspects of projects, but staff need to be persuaded to fill them in conscientiously.

Many phrases which refer to **informal monitoring** spring to mind: keeping one's ear to the ground, management by walking about, open door policy. All of these indicate that the manager has an awareness of what is going on 'at the coal face'. Project managers need to ensure that they have a way of maintaining good informal lines of communication with all project personnel. The advantage of this approach is that progress and problems will be communicated more quickly than with more formal methods. However, a pitfall to avoid is the alienation of team members by over-supervision.

## 3.4 Applying control

There is no point in monitoring unless **control** is going to be applied. This is done through the process of the **reporting cycle**. What normally happens is that the monitoring processes described above identify some shortfalls in the progress of the project. To remedy shortfalls, control needs to be applied to the project to bring it back on course. For example, staff may be transferred from non-critical work to critical activities that have fallen behind.

The reporting cycle defined in the project initiation document identifies who should be producing progress reports, with what frequency and also to whom they should be sent. Remember that reporting is an overhead. Reports should therefore be concise, contain only relevant information and be circulated only to those who need to see them. However, it should be noted that more concise reports require more effort by the writer in order to save the time of the readers. As someone once apologized: 'I am sorry this report is so long: I didn't have time to write a shorter one.'

The reporting structure refers to the people involved in a project at different management levels. This structure should be determined before the project starts and should be laid down in the appropriate documentation, usually the project initiation document (see Chapter 1).

Generally progress reporting starts at the level at which work is actually done and progresses up through a hierarchy. In an IT context this means team leaders gather progress information from their team and report up to their project manager who then reports to the group that has been entrusted with overall responsibility for the project (the project board or steering committee). This group would have representatives of the managers of the development team, the users and the project sponsor. They, not the project manager, would have the authority to change the objectives of the project. For example, they could allocate more resources to the project or reduce the scope of what is going to be delivered. The report to the project board or steering committee is sometimes referred to as a **highlight report**. The intervals at which the reports are produced and the topics they report on need to be tailored to the specific requirements of the recipients and the importance of the information to be conveyed. It is important to obtain formal agreement to the reporting procedures from all the parties involved.

## 3.5   Purpose and types of reporting meetings

### 3.5.1   Team meetings

These are usually attended by team members, the team leader and possibly the project manager. A weekly frequency is usually appropriate. A report from the team leader to the project manager would be prepared.

A typical agenda would include the following:

- each individual team member's progress against their plans;

- reasons for variances;

- expected progress;

- current problems or **issues**;

- possible future problems – this may involve reviewing the risks that have been recorded in the **project risk register** (see Chapter 7) that could affect this part of the project.

It is important that all those attending have a reason for attending and a contribution to make. These meetings are often referred to as **checkpoint** meetings and the progress report produced in this case is a **checkpoint report**. As issues are identified they may be recorded in an **issues log** which will be updated as they are resolved.

## 3.5.2 Project board meetings

It will be recalled from Chapter 1 that project boards may have different names in different organizations, for example steering committees or project management boards. These meetings will be attended by board members and the project manager, with secretarial support, perhaps provided by a project support office. The structure and responsibilities of project boards, and the project support office, are covered in Chapter 8.

The frequency of meetings may well be monthly but the exact timing would depend on the project size: larger projects may have fewer and less frequent top-level meetings, but more meetings of managers at intermediate levels. Meetings could well be timed to coincide with significant project events such as the completion of a particular project phase or stage or possibly other external triggers such as requirements for financial approval.

Items for the agenda would be similar to that for team meetings. A report from the project manager to the board would be written and circulated prior to the meeting. This **highlight report** would be condensed from the checkpoint reports produced by the preceding team meetings. The board would have to exercise its authority to make decisions relating to any necessary corrective action arising from progress information. This would be fed back down the reporting chain and thus complete the reporting cycle.

The highlight report may contain the following information:

- details of the progress of the project against the plan;
- current milestones achieved;
- deliverables completed;
- resource usage;
- reasons for any deviation from the plan;
- new issues and unresolved issues;
- changes to risk assessments;
- current problems;
- any anticipated future problems/risks;
- plans for the next period and products to be delivered;
- anticipated completion of current products and activities;
- graphical representations of progress information.

## 3.5.3 Programme board/steering committee meetings

Organizations sometimes group projects into **programmes,** where a number of projects all contribute to a set of over-arching objectives (see Chapter 8). In these cases a programme board may be set up, to which individual project boards would report. These boards would have less frequent, less detailed meetings related to programme management but with essentially a similar agenda to those of the project boards. These would have more of a business focus than a project focus.

## 3.6   Taking corrective action

Here we will examine how corrective action can be applied in a considered controlled way. The project manager's role is to manage on a day to day basis, applying minor corrections as required. However, corrections of a more major nature will need referring to higher authority.

This is the reason for allocating a **tolerance** within which the project manager has authority to make changes or apply corrective action. For example, you, as a project manager, may be allocated 10% tolerance on time and cost on a project worth £100,000 and lasting 25 weeks. You would be authorized to agree to changes worth £10,000 or an overrun of two and a half weeks (see Chapter 4). If, on the other hand, a situation was arrived at where the project was expected to exceed the two-and-a half-week overrun you would be in an **exception** situation. In this case, you would need to prepare an **exception report** for submission to a special meeting of the project board. This exception report would probably outline a number of **options** designed to correct the overrun and the board would come to a decision on the best way to proceed.

Tolerance and **contingency** pools are sometimes distinguished. Tolerances can be assigned to individual activities within the project. The contingency pool is a set of resources that is controlled by the project manager and can be allocated at the discretion of the project manager where additional resources are needed. However, if the contingency is used to buy resources in one place it means that there will be less money for other emergencies. These resources may be augmented where activities are completed early and so free up resources. The important point to note here is that where activities can be completed early, the opportunity should be seized as this will release staff to deal with unanticipated delays elsewhere.

An exception report typically includes the following information:

- background;
- reasons why the exception arose;
- options;
- risks;
- **exception plans** showing how the project plans need to be amended in order to implement the suggested options;
- amended business case;
- recommendation.

When the project board (or equivalent) members scrutinize the exception report, they will be particularly concerned to ensure that the business case for the project will be preserved, that is, that the costs of the project will not exceed the value of its eventual benefits. If the board are satisfied with the exception report, the project manager is given authority to proceed using the chosen option and its associated exception plan.

Where monitoring reveals a shortfall in the expected progress, control is applied to bring the project back on track. There are a number of standard control strategies, which may or may not require an exception report:

- **Work harder, longer or faster**

  This is the most obvious approach and often appeals to more aggressive managers. However, although it may work to solve a short-term problem or to meet a critical deadline, it will fail if used too much. Staff will become tired, stressed, and then demotivated.

● **Re-plan**

There may be some scope for re-planning. Although some project activities have consumed more staff effort or taken longer to complete than planned, others may have taken less effort or time. Adjustment of resources may therefore be possible but this may be interpreted to mean that the project did not start with an optimum plan.

● **Extend the timescale**

This is another frequently adopted option. However, there is usually a deadline, to which changes require negotiation, usually through the exception reporting process described above. Extending the deadline is often seen as 'weak management' or allowing the project to get 'out of control'. However it can be the most sensible option.

● **Increase resources**

This needs to be carefully considered, particularly in the development stages of an IT project. Resources in this context means people; adding more does not usually increase productivity and often decreases it. The introduction of more staff involves a period of induction while they familiarize themselves with the work. The current staff will inevitably be involved in this process and the overall effect could even be to delay progress. An exception report and plan would be needed if the additional costs would go beyond the tolerances that have been agreed for the project.

● **Reduce the project scope**

This is an attractive option which also needs negotiation with management and drafting on an exception plan. Deliverables may be removed from the plan or delayed until after the planned project end date. This does not affect the originally planned cost or duration of the project.

● **Terminate the project**

If no acceptable alternative can be found this may be the only remaining sensible action. Terminating the project would be justified if it is clear that the remaining costs of the project will exceed the projected value of its benefits when delivered.

For each of the options in the list above, identify in general terms the possible impact on the project's business case.

*Activity 3.2*

## 3.7 Graphical representation of progress information

### 3.7.1 Gantt chart

Progress information can be shown on a Gantt chart by shading completed or partly completed activities. Figure 3.1 shows the Gantt chart that was produced in Chapter 2, updated to show the actual situation at the end of week 3 of the project. At this point the following information could be reported:

| Table 3.1 | Reporting Gantt chart information |
|-----------|-----------------------------------|

| Activity | Status |
|----------|--------|
| B. allocate room | only took one week to complete, not two |
| C. write/test software | seems to be going well – reported as 50% completed |
| D. draft office procedures | started two weeks late – 25% complete |
| E. install office furniture | not started – awaiting delivery from supplier |

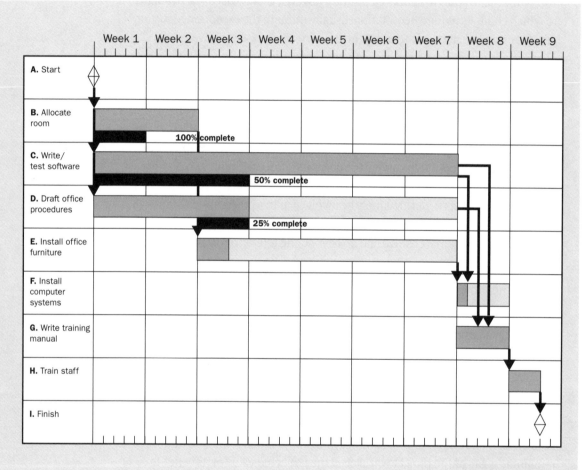

Fig. 3.1
A Gantt chart that has been updated with actual progress

In week 4, the following actions take place:

(a) Activity C write/test software – because of technical difficulties, no product can be signed off;

(b) Activity D draft office procedures – work continues at the same rate as the previous week;

(c) Activity E install office furniture – this is delivered and installed.

Update the Gantt chart in Figure 3.1 to take account of these changes.

*Activity 3.3*

## 3.7.2 Cumulative resource chart

A cumulative resource chart can be used to present resource usage details (see Figure 3.2). It is sometimes called an S-curve chart, because the pattern of activity is that a project starts with a relatively low number of staff at the planning and requirements gathering stage. As the project progresses, more and more staff are needed as the development and implementation activities multiply. The demand for staff then decreases as the implementation proceeds – for example, software developers are not needed full-time once they have constructed their software components. This pattern of activity, a low demand for staff time which gradually builds up and then declines, leads to a cumulative resources chart where the lines seem to have an approximate S-shape.

These charts normally have two sets of data points: one showing the expenditure that was planned and the other the actual expenditure, for comparison.

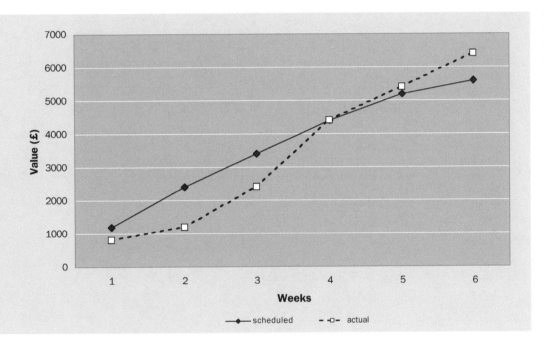

Fig. 3.2
A cumulative resource chart

Typically, management is provided with a visible representation of a project's progress by S-curve charts that track accumulated costs over time, for comparison with the original estimates.

Figure 3.2 shows that for most of the project we were under-spending, but that there has recently been a surge in expenditure, so that currently we have spent more than planned. However, we do not know whether this is due to poor productivity, or whether we have actually produced more than was scheduled – work may have been completed early, leading to some expenditure being incurred earlier as well.

The traditional S-curve chart does not show any of the following:

● whether the project is ahead or behind schedule;

● whether the project is getting value for money;

● whether problems are over or just beginning.

## 3.7.3 Earned value analysis

If we plot a third line, the earned value, then we can see if we are ahead or behind time, and above or below budget. **Earned value analysis (EVA)** shows the budget that was originally allocated to items of work that have been completed. When the work is finished, we can say that this value has been 'earned'.

For example, let us assume that in the project schedule shown in Figure 3.1, it is possible to identify products of Activity C that are complete and can be signed off at the end of each of the seven weeks of the activity. If one developer is allocated to the activity at a cost of £400 a week, Activity C is given a budget of £400 × 7, that is £2,800. At the end of week 3, £400 × 3 or £1,200 of the budget has been used. However we may be in the fortunate position at the end of week 3 to sign off work originally planned to be produced in week 4. This means that the earned value of the task at this time would be £400 × 4, that is £1,600. Unfortunately it is often the case in reality that the earned value is less than planned.

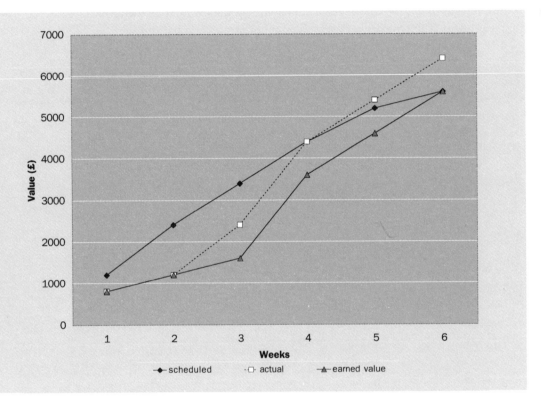

**Fig. 3.3**
Earned value graph

In Figure 3.3, a line showing earned value has been added to the cumulative resource graph. This shows, for example, that at the end of week 6, the project is on schedule and has completed the work that was planned. However, it can be seen that expenditure is greater than planned. This may be a case where the project manager got the project back on schedule by buying in overtime.

Please note that candidates for the ISEB foundation certificate are not expected to know the details of how earned value is calculated.

## Sample questions

**1.** Which of the following would you most expect to see in a routine report from a project manager to a project board?

(a)   Costs and benefits

(b)   Progress against plan

(c)   Configuration status information

(d)   Current activity

**2.** Which of the following is **not** involved in collecting progress information?

(a) Team progress meetings

(b) Timesheets

(c) Comparing planned and actual costs

(d) Informal monitoring

**3.** Which of the following would usually give rise to an exception report?

(a) A new issue being raised

(b) A proposal to make a change to a deliverable

(c) The unexpected loss of a key team member

(d) Project tolerance being exceeded

**4.** What is the purpose of earned value analysis?

(a) Assessing progress

(b) Collecting progress information

(c) Estimating the required effort

(d) Calculating benefits

# Answers to sample questions

1. (b)    2. (c)    3. (d)    4 (a)

# Pointers for activities

## Activity 3.1

(a) If we keep to the original planned training delivery rate of one boatyard a day, 17 days (that is, about three to four weeks) are needed to complete training, as there are 17 boatyards left. However, if we decide that the experience of the first week shows that the original delivery rate was unrealistic, then we may project that three boatyards can be visited in a week. This would lead to between five and six weeks being needed to complete training.

(b)(i) In this scenario, the reason for only three boatyards being visited was simply a late start rather than a low delivery rate. The estimate of 17 days would seem to be reasonable.

(ii) The reason for lateness here was a lower delivery rate. However as the training programme proceeds, the delivery rate should improve as the nearer boatyards are dealt with, and journey times get shorter. Revising the planned time to five to six weeks would be premature at this point.

This activity should illustrate how informal information gathering can help interpret more formal reports.

## Activity 3.2

The basic question with respect to the business case is whether the value of the benefits of the delivered system will exceed the costs of development, implementation and operation. The points below are merely pointers and many other possible factors could be discussed.

- The key factor about working longer, harder or faster is whether overtime is paid or not. If no overtime payments are made, then costs will not increase. If payment is made then clearly costs will increase and this could have a detrimental effect on the business case. There is a risk in this situation that the benefit to the end user could be decreased through a poorer quality product.

- As existing resources are rescheduled and the overall project duration is not affected, re-planning should not affect the business case.

- Extending the deadline often, but not always, implies additional staff costs. However, using a smaller number of staff for longer can be more efficient than using a large number for a intensive period (see Chapter 2). The benefits to the users will be reduced as they will have to wait longer to receive them (see the discussion on net present value in Chapter 1).

- Increasing resources will clearly increase costs.

- Although costs will be unaffected by reducing the project scope, the benefits that the users receive will be reduced. They may have to pay out for additional development at a later date to obtain the deferred system features.

## Activity 3.3

See Figure 3.4.

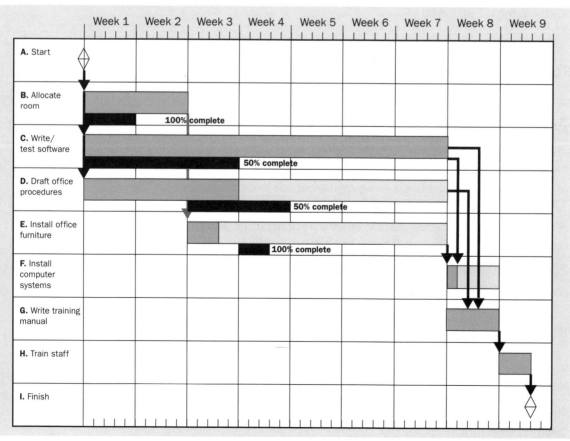

Fig. 3.4
A Gantt chart that has been updated with actual progress up to week 4

04

# Change control and configuration management

**LEARNING OUTCOMES**

When you have completed this chapter you should be able to demonstrate an understanding of the following:

▶ the reasons for change and configuration management;

▶ change control procedures:

- the role of the change control board;

- the generation, evaluation and authorization of change requests;

▶ configuration management:

- purpose and procedures;

- the identification of configuration items;

- product baselines;

- the content and use of configuration management databases.

## 4.1 Introduction

Many events during a project result in changes to the planned work. The project manager's skill lies in controlling those changes so that minimum disruption is caused to the planned objectives of time, cost and quality. The purpose of **change management** is to ensure that modifications to any aspect of the project are only accepted after a formal review of their impact upon the project as a whole.

The procedures for managing change should be established at the beginning of the project. Many large organizations will already have a process in place. However, Canal Dreams is a relatively new company which has had few, if any, major IT development projects. In this case, a change control system would need to be devised and implemented, especially if outside contractors are going to be writing the software. An obvious step in any change management process is to make allowance for it within cost and time plans. This allowance may be part of the tolerances delegated to the project manager (see Chapter 3). The project manager's skill will then be exercised in ensuring that the effect of any change does not exceed these allowances. Where the allowance for changes is exceeded, a new project plan (the exception plan mentioned in Section 3.6) would need to be drafted and approved by the steering committee or project board.

Change management, at some level, should be applied to all changes, whether they arise from a change to user requirements or a change due to design errors. Change management requires participation by the users and the developers, guided by the project manager. For every change the user must decide whether it is essential, desirable or optional. It is the project manager's function to identify the cost, time and quality impact, and to assess the feasibility of the change. Once account has been taken of the users and the project manager's advice, if the decision is to go ahead with the change, it is the responsibility of the project manager to implement the change.

## 4.2   Definition of change

In Chapters 2 and 3, we introduced the idea of the project as a sequence of activities, each of which take certain inputs and use them to produce outputs. The outputs from one process may be inputs to other processes. For example, a specification could be an input to a process that creates a system design (that is, the format of the user interfaces with the system) which will then be implemented as code. During the process, the proposed interface designs may change rapidly as the designers try out different designs and the users give feedback on them. During this process, a formal change process is not required. At some point, however, a decision needs to be taken that the design is now in a satisfactory state to be used as a basis for constructing the system. At this point the design will need to be frozen; in other words, the design is **baselined**. If changes are made to the application after it has been baselined, then rigorous change control is needed because of the potential impact on subsequent phases.

Any change can also be seen as a potential change to a **baselined plan**. A baselined plan may be regarded as an agreed plan against which variations will be measured. For projects, the baselines include:

- the agreed scope and quality of work;
- the agreed schedule;
- the agreed cost.

As noted in Chapter 3, these will tend to be interdependent. For example, if the scope of the work is expanded, then the cost will also have to be increased. Cost or the scheduled duration could be reduced by decreasing the functionality in the IT application that is to be delivered.

Variations on these baselines can be categorized as follows:

- changes of scope;

  This type of variation generally originates outside the project, usually by the user changing the requirements or by the cost or time constraints of the project changing. In the case of the Canal Dreams scenario the users may, for example, add a requirement that when booking a boat the client should also be able to purchase holiday insurance. On the other hand, a reduction in the project budget may mean that some system features that were planned must be dropped.

- development changes;

  This type of variation originates within the project and includes changes which are routinely carried out as part of the normal process of developing and refining a product. Typically, this can be something as simple as an adjustment to a screen layout.

- faults;

    This type of variation also originates within the project and is caused by the project team making an error. For example, errors in coding may result in erroneous results being produced.

Some discretion may be exercised with regard to accepting or rejecting scope and development changes but changes due to design errors will normally be obligatory since the system may not work satisfactorily unless they are corrected.

**In order to encourage the UK canal holiday industry, the government decides that VAT on canal holiday bookings will be zero-rated with immediate effect. The management of Canal Dreams calculate that this could increase the demand for bookings by 300%. At the same time, the government introduces a special tax on canal holiday insurance which has to be accounted for separately on invoices. When considering the implications of changes, the project team realize that although holiday insurance was included in the original requirement, it has been missed out from the system design. What effect may these changes have on the project?**

*Activity 4.1*

## 4.3 Change management roles and responsibilities

It is important to clarify the various roles and responsibilities within change management. We are going to use a very specific set of terms here to identify and explain the various roles and responsibilities. In a real project environment, it is very unlikely that these precise terms will be used. However, there should be people who carry out the following roles, whatever they may be called:

- project manager;

    The project manager oversees the process and ensures that all **requests for change** (RFCs) are handled appropriately. In most cases, the project manager also has the role of change manager.

    The project manager is concerned with ensuring that the users continue to know what to expect when their system is implemented. The project manager also controls the scope of the system to be developed: additional features will require more effort and increase costs. When the project sponsor agrees to additional features, adjustments may need to be made to the contractual price for the project. The project manager also takes note of any informal changes made outside the process. Informal changes are often discovered during project reviews and a retrospective RFC form may be completed in order to reconstitute the flow of the change procedure.

- change requestor;

    The **change requestor** recognizes a need for a change to the project and formally communicates this requirement to the change manager, by completing the RFC form.

- change manager;

    The **change manager** (likely to be the project manager) is responsible for logging RFCs in the **change register**. The change manager also decides whether or not a feasibility study is required for the change. Where this is desirable, for example because the extent of the impact of the change is uncertain, one or more people will be assigned to carry out the study.

● change feasibility group;

The **change feasibility group** investigates the feasibility of a proposed change. It is responsible for researching how the change may be implemented and assessing the costs, benefits and impact of each option. Its findings are documented in a feasibility study report. If the staff who are carrying out the feasibility study are also project team members, then there is a risk that project activities may be delayed.

● change control board;

The **change control board** (CCB) decides whether to accept or reject the RFC forwarded by the change manager. The CCB is responsible for:

- reviewing all RFCs forwarded by the change manager;
- approving or rejecting each RFC based on its merits;
- resolving change conflict, where several changes overlap;
- resolving problems that may arise from any change;
- approving the change implementation timetable.

● change implementation group.

The **change implementation group** carries out the change. It is usually made up of staff from the project team.

## 4.4    The change management process

We have already outlined some of the processes involved in implementing a change:

● submission and receipt of change requests;

● review and logging of change requests;

● determination of the feasibility of change requests;

● approval of change requests (or rejection or 'putting on hold');

● implementation and closure of change requests.

### 4.4.1    Submit request for change

In the Canal Dreams project, the project manager has also been allocated the role of change manager. A former booking clerk at one of the boatyards who has taken up a co-ordinating role in the central finance office has been selected to act as change requestor. Requests for change can come from anyone and the change requestor, in consultation with colleagues decides whether the requested change is desirable from the users' point of view. If they decide that there is a genuine need, a request for a change is submitted to the change manager. The RFC provides a summary of the change required, including:

● a description of the change needed;

● the reasons for change, including the business implications;

● the benefits of change;

● supporting documentation.

### 4.4.2 Review request for change

The change manager reviews the RFC and determines whether or not a full feasibility study/impact analysis is required in order for the CCB to assess the full impact of the change. In the Canal Dreams project, one request was to add the sale of holiday insurance to the booking system. On receiving the RFC, the change manager sees that it is difficult to assess the scope of the change as some of the details of the requirement are still unclear. The impact of the change on the existing design also needs to carefully considered. The change manager opens an RFC entry in the change register and records that a feasibility study is required.

### 4.4.3 Assess feasibility of change

Once the change has been logged, an assessment must be made of its feasibility and impact upon the project in terms of time, cost and quality. At the simplest level, a team member may assess the impact of the change in a very informal manner; for major changes, the feasibility study may involve several people over a period of time. All change options will be investigated and presented accordingly. The change feasibility study will culminate in definition of the:

- change requirements;
- change options;
- costs and benefits;
- risks and issues;
- change impact;
- recommendations and plan.

A quality review of the feasibility study is then performed in order to ensure that it has been conducted as requested and the final deliverable is approved – ready for release to the CCB. All change documentation is then collated by the change manager and submitted to the CCB for final review.

The feasibility study itself carries a cost and the practice is sometimes that the project manager records these costs in the change register. These costs may well increase the budget. For external client projects, the feasibility study may be an additional service which has to be paid for by the customer.

In the Canal Dreams project, two staff are assigned to look at the requested change. One is a business analyst who should understand the nature of the business and the other is an experienced developer who will be able to judge the technical impact of the requested change. After discussing the matter with the user representatives they find that the change requires two additional input fields on the booking screen, two additional data items on one of the tables in the database and some minor changes to some of the printed outputs from the system. They estimate that the changes would require three additional staff-days of effort.

### 4.4.4 Approve request for change

A project may have a range of levels at which changes can be reviewed and approved:

- Team leaders may be allowed to accept changes that will not require additional resources and which do not affect other baselined products.
- The project manager may be allowed to decide upon changes which have a

minor impact on project objectives within a tolerance level which has been agreed with the steering committee or project board.

● The CCB decides upon changes which will have a larger impact upon project objectives, but are constrained by any limitations on the budget available for changes.

Whatever the level of review, the change needs to be recorded and reported.

The CCB will do one of the following:

● reject the change;

● request more information related to the change;

● approve the change as requested;

● approve the change subject to specified conditions;

● put the change on hold.

The CCB will need to take account of the overall profile of possible changes. A large number of minor changes could have an overall effect out of all proportion to their individual significance. Approved changes will necessitate revisions to the schedule and cost plan. The CCB should prioritize changes so that those that are essential are carried out first, while non-essential changes are delayed until they can be made with least impact on work schedules.

In the Canal Dreams project, the holiday insurance change is only one of several changes that have been requested. The CCB has a budget of only 10 staff-days left to allocate, but has requests that in total need 15 days. The CCB may accept changes requiring up to 6 days effort, including the change to incorporate the holiday insurance requirements.

### 4.4.5　Implement request for change

This involves the complete implementation of the change, including any additional testing that may be required. On completion, the change would be signed off in the change register. Where the additional work has been carried out by an external supplier, additional invoices for the additional work would be raised by the supplier. Additional payments would not be made, however, where the change was to correct an error made by the supplier.

## 4.5　Configuration management

The change control system described above is needed to ensure that an endless sequence of changes does not undermine the business case for the project, for example, by increasing costs so that they exceed the value of the benefits that the project will produce or by extending the time needed for development and implementation and thus reducing the benefits. Once a change has been agreed, there are further problems of ensuring that all documents and other products of the project are modified to reflect the change.

**What deliverables of a project may be affected by the change to the Canal Dreams specification to allow for holiday insurance to be recorded when a customer books a boat?**

*Activity 4.2*

A project has a wealth of documents that relate to each phase of the project, along with software objects such as database structures and code segments. A very basic need is therefore for a central **project repository** or **library** where master copies of all documents and software objects are stored. Another basic requirement is that there should be a system of **version numbers** for all products so that successive baselines can be identified. These requirements mean that it is essential that there is one or more people involved in the project who have a role that is variously called **project** or **configuration librarian** or **configuration manager**. Part of that role is to make sure that all project products are controlled so that, for example, we can make sure that all software developers who are working on components that pass information are working from the latest specifications for those components.

Configuration management has three major elements:

- configuration item identification;
- configuration status accounting;
- configuration control.

## 4.5.1 Configuration item identification

The items which will be subject to the **configuration management system** (CMS) need to be identified. Typically these are baselined specifications, design documents, software components, operational and support documentation, and key planning documents such as schedules and budgets. Other items such as IT equipment may also be subject to configuration management. These items will be defined as **configuration items** (CIs) and their details will be recorded in a **configuration management database** (CMDB). Among the details recorded in the CMDB for a configuration item are:

- a CI reference number;
- its current status;
- its version number;
- any larger configurations of which it is a part;
- any components that it has;
- other products that it is derived from;
- other products that are derived from it.

## 4.5.2 Configuration status accounting

After a change to a CI has been agreed, the project librarian sets the status of the CI accordingly. This process is called configuration status accounting and it maintains a continuous record of the status of the individual items which make up the system.

## 4.5.3 Configuration control

Configuration control ensures that due account is taken of the status of each CI. For example, when recording the change to add holiday insurance to the boat booking transaction, the configuration librarian may access the CMDB to see the current status of the software component. The librarian may find that the software component is already booked out to a software developer who is implementing a different approved change to the module. If the librarian were to release a copy of the baselined code to a second developer to add holiday insurance then that would create two different versions of the same software. The new change may be given to the developer already working on the component or there may need to be a delay while the first change is completed.

When a developer is happy that all the work associated with a change is complete, the new version of the software is passed to the librarian. The librarian then records the CI as having a new version number and as being ready for acceptance testing. The former version of the software is retained in case there is a need to 'fall back' to it, if the new version turns out to be problematic.

## Sample questions

**1.** The change control board should consist of:

(a)    representatives of the key stakeholders in the project;

(b)    the project manager and team leader;

(c)    the people on the project board;

(d)    the project support office.

**2.** If there are doubts about the projected costs of a proposed change, it would be the responsibility of which of the following to investigate?

(a)    the change requestor;

(b)    the change manager;

(c)    the change feasibility group;

(d)    the change control board.

**3.** As a documented procedure, what is the purpose of configuration management?

(a)    to ensure the project remains within budget;

(b)    to identify and document functional characteristics of a system;

(c)    to record and report changes and their implementation status;

(d)    to verify conformance to requirements.

## Answers to sample questions

1. (a)    2. (c)    3. (c)

## Pointers for activities

### Activity 4.1

The changes may have the following effects on the project:

● The change to the rate of VAT should **not** involve changing the application as there should already be a system function that allows VAT rates to be changed.

● The potential increase in bookings would not change the functional requirements, but would probably change the quality or 'non-functional' requirements. For example, the database would need to be able to hold more records. The equipment needed to run the application may need to be upgraded (this is a change to the scope of the project).

● The statutory change to accommodate holiday insurance tax could well mean changes to input screens and report layouts (this is a change to the scope of the project).

● The design not including functions to deal with holiday insurance is a straightforward fault and the project team must correct it.

## Activity 4.2

Among the many deliverables that may be affected are:

● the interface design – screens and report layouts;

● software components;

● the database structure;

● test data and expected results;

● user manuals;

● training materials.

# Quality

---

### LEARNING OUTCOMES

When you have completed this chapter you should be able to demonstrate an understanding of the following:

- ▶ definitions of 'quality';
- ▶ quality control and quality assurance;
- ▶ measurement of quality;
- ▶ detection of defects during the project life cycle;
- ▶ quality procedures: entry, process and exit requirements;
- ▶ defect removal processes, including testing and reviews;
- ▶ types of testing (including unit, integration, user acceptance and regression testing);
- ▶ the inspection process and peer reviews;
- ▶ the principles of ISO 9001:2000 quality management systems;
- ▶ evaluation of suppliers.

---

## 5.1 Introduction

The critical aspect of quality as far as projects are concerned is to provide the customers with the systems that they need and which meet their requirements at a price they can afford. In order to achieve this, quality has to be built into a product. It cannot be added to a product after it has been produced, without great difficulty and cost. It is like trying to alter the foundations of a building once it is complete.

To build in quality requires a commitment from all parties involved in the project, from the project sponsor through all the levels of management, the technical staff, customers and users and the clerical support staff. Back in 1979, Philip Crosby wrote a book in which this is the opening paragraph:

'Quality is free. It's not a gift, but it is free. What costs money are the unquality things – all the actions that involve not doing the job right in the first place.'

> Philip B Crosby, *Quality is free*, Mentor, 1980.

Nothing has changed. There is, as explained in Chapter 1, a relationship between quality, cost and time. The level of quality required will have an impact on both time and cost, but – and it is a very important but – the more effort that goes into quality at the beginning of a project the less the expenditure at the end of the project correcting faults.

## 5.2 Definitions of quality

A dictionary definition of quality is 'a degree or level of excellence' as in the phrase 'high-quality goods'. This is relevant to projects as it is hoped that all products are of a high quality. But the definition is subjective: for example, when comparing cars, people would disagree between the quality of different makes. Another definition is 'conformance to standard'. Again this has an application to projects as within the project process there will be certain standards to which those developing the system ought to conform. However, this does not tell the whole story. The generally accepted definition which should apply to all projects is that the deliverables should be **'fit for purpose'**. Again referring to cars, a Rolls Royce may not be the best vehicle if your main need is to get to work through heavily congested traffic on time: a scooter or motorbike might be more effective. The international standard ISO 8402 sets out the formal definition of this meaning of quality as:

'The totality of features and characteristics of a product or service which bear on its ability to satisfy stated or implied needs.'

Looking at just one aspect of this 'fit for purpose' definition – reliability – it can be shown how the concept applies. If the Canal Dreams system fails after it has been delivered it would be annoying but is unlikely to be life threatening. However, if the control systems in an aircraft fail when it is in flight, that would be disastrous. The effort, and hence the cost, that has to be put into making sure that the aircraft system does not fail would be considerably more than that one would want to put into the Canal Dreams system. Hence quality is a variable depending on the type of system under development and the amount of money the customer is prepared to pay. To a large extent, it is the customer who must decide the level or degree of quality that should be built into a system. Those associated with the project from a technical angle must advise the customer on the benefits of a well-engineered system but in the end it is the person paying the bill who should call the tune. However, a supplier would have a professional and legal commitment to ensure that the systems they produce are safe.

Because software systems are becoming ever more complex, it is impossible to ensure that the software will never fail. In addition, there is the hardware and infrastructure which might also be subject to failure. When considering quality it is necessary to discuss the ways in which the systems under development will behave in the event of various types of failure as this will directly influence the development process and the ways in which quality is measured. For example, consider a control system for a nuclear power plant. It will be designed to 'fail safe', that is, in the event of a systems failure it will revert to a safe state, for instance it will close down. Obviously such a requirement must be specified as part of the quality criteria for the system. It should be noted that where something is inherently dangerous, as in this example of the nuclear power plant, the developers would have a duty of care, not just to the project sponsor, but to the world at large.

## 5.3 Quality characteristics

The definition 'fitness for purpose' needs elaborating so that it can be applied practically. Another international standard, ISO 9126-1:2001, seeks to define a set of standard characteristics by which software quality can be measured. It specifies six characteristics at a high level:

- **Functionality**: does the system as delivered meet the functional requirements of the user? Meeting user expectations is something more than just meeting a specification.

- **Reliability**: how often does the system break down and how long does it take to put right? Are the results it produces consistent?

- **Usability**: is it straightforward to use? How much training is required for someone to be able to use it?

- **Efficiency**: how much physical resource is required to operate the system?

- **Maintainability**: all software is subject to change; can this software application be modified easily and without introducing unexpected errors?

- **Portability**: how easy is it to take the software from the particular platform on which it was developed to another environment?

The standard itself goes into a lot more detail for each heading, but these top-level qualities are useful guides, particularly when trying to establish measurable quality criteria.

## 5.4 Quality criteria

What are quality criteria? To ensure that quality is built into a product the level of quality required has to be specified at the beginning, before the product is developed. In Chapter 2, the concept of a **product description** was introduced. Each product description has the same structure and includes a section headed **quality criteria**. The accurate completion of this section will enable project team staff later on in the development cycle to determine whether the product is fit for purpose, by checking it meets the quality criteria specified.

In order to achieve this, the criteria themselves have to have three qualities:

- They must be specific.

- They must be measurable.

- They must be achievable.

As an example, the aircraft control system has a specific requirement that the product must never fail. That is quite clear. It is also measurable: by testing the product for an exhaustive amount of time until the product fails it can be demonstrated that the requirement has **not** been met. In this case, we can prove when the quality criterion has not been met, but we cannot prove that it will not fail at some point in the future. Hence, it is not achievable! In management, SMART objectives are used. This mnemonic stands for:

- Specific;

- Measurable;

- Achievable;

- Relevant;

- Time-constrained.

The same mnemonic can be used in relation to quality criteria.

An organization should develop standards for things like product descriptions and these will form part of the quality procedures for that organization.

Refer back to the discussion about product definitions (Section 2.2.1). Draw up a list of quality criteria for a product definition document (not for a product). Add any other headings that you consider should be standard for all product descriptions. Specify how the criteria may be measured.

*Activity 5.1*

The following are examples of good and bad quality criteria.

*Activity 5.2*

● Screen layouts should be consistent.

● Screens should be user friendly.

● The system should be able to handle 50 transactions.

● The system should allow for 20 users at any one time without degradation.

● The response time should not be longer than three seconds.

Identify the good and bad criteria and say why they are in that category.

Maintainability is defined as a quality characteristic. How can it be measured?

*Activity 5.3*

How can the reliability of a system be measured?

*Activity 5.4*

## 5.5   Quality control versus quality assurance

Having introduced the concept of quality criteria and how they should be specified, it is time to step back and discuss the framework within which quality activities take place. In an ideal environment the project will take place in an organization which is committed to quality. This implies that standards will already be in place for certain activities. If this is not the case, then the appropriate parts of the project organization, for example the project board or the project manager, will have to be more active in setting the framework for the quality of the project.

The quality framework is called the **quality management system (QMS)**. It may be based on the ISO 9000 series of standards, in particular ISO 9001:2000 (see Section 5.10). Within the QMS, there will be a quality strategy and quality assurance processes that provide a general framework for projects within the organization. They have to be reviewed for each new project and, if necessary, modified to meet the specific requirements of that project.

A **quality strategy** defines the QMS and includes:

● procedures and standards for creating a project **quality plan;**

● a definition of **quality criteria**;

● **quality control** procedures;

● **quality assurance** procedures;

- a statement of compliance with or allowed deviation from industry standards;

- **acceptance criteria;**

- an allocation of responsibilities for defining and or undertaking the above activities.

Thus the methods by which quality is to be controlled are specified in the strategy. The methods for exercising control are discussed in detail later, but it can be said now that each method can be classed as some form of review. Quality assurance stands alongside reviews but is external to it. Quality assurance is like an audit. It is designed to confirm that proper procedures are in place and that they have been applied correctly. The difference between the two can be illustrated by the following example. A metal rod must be 15mm in diameter ($\pm$ 0.1mm). Control is achieved by passing each rod through a measuring device which triggers rejection if the rod is out of tolerance. Assurance is exercised by checking that the measuring device is accurate and that all rods go through it.

Quality assurance applies to projects in the following way. For each component, quality criteria are specified. The setting of quality criteria is in effect the first part of quality control for without the criteria no control can take place. A control process is undertaken to ensure that the criteria have been met. This is followed, at an appropriate time, by an assurance process which confirms that the agreed procedures have been followed and that all products have undergone the necessary checks.

You will recall from Chapter 1 that the systems development life cycle has a number of stages, typically:

- initiation;

- feasibility study;

- project set-up;

- requirements analysis and specification;

- systems design;

- detailed design;

- construction;

- testing;

- implementation;

- review.

Each stage contains several quality control processes, in the form of some sort of review, and a quality assurance process takes place at the end of each stage. If proper quality control has been lacking, then corrective action may be mandated. The quality control and, if necessary, the quality assurance processes are repeated until the quality criteria are met. (Quality assurance processes may take place more often if the project requires it.)

## 5.6 Quality plan

The production of a **quality plan** is vital to the success of a project. It specifies the particular standards that apply to the project. They could be existing standards, defining the content and layout of product descriptions, for example; modifications to existing standards because of some variation specific to the current project; or completely new, as the project goes into territory not previously explored. The standards may derive from industry standards or be developed internally. Normally most standards come from existing processes and procedures in the company. If not, then an early task of the project manager is to ensure that the standards are specified.

The plan also specifies how, when and by whom the quality control activities should be undertaken and the quality assurance processes to be followed and who will carry them out. It may also include configuration management and change control procedures (see Chapter 4).

The quality plan itself is subject to quality control and quality assurance processes. Quality control would normally be an integral part of the project team's work. The quality assurance activities, on the other hand, are usually carried out by people outside the project team who report directly to the project board. This separation of responsibilities helps to ensure that the process is transparent and reduces possible conflicts of interest.

## 5.7    Detecting defects

### 5.7.1    The V Model

The V Model is a useful model of the systems development process (see Figure 5.1) in which the solid lines represent the forward progress of the project and the dotted lines represent the way in which quality control is exercised.

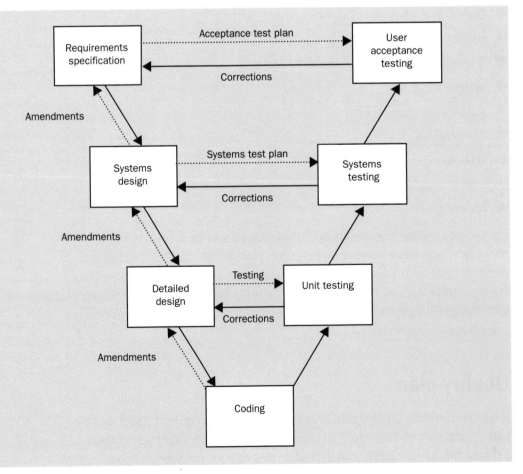

Fig. 5.1
A simplified V model

There are two quality control processes at work: one between stages and the other across the V. For example, the requirements specification is a key document. It should contain a specification of all the functions and quality attributes that the customer requires in the system. In Chapter 1, the process of creating this specification included creating an acceptance test plan. As part of the acceptance criteria, the acceptance test plan should show how the requirements are going to be measured in the final system. Using the acceptance test plan, user acceptance testing can be undertaken to demonstrate that the system does indeed meet the requirements identified.

Systems design follows requirements specification and is joined to it by an arrow. As part of the quality control process for systems design it is necessary to check that everything specified in the requirements specification has been incorporated into the design. If not, then the design is incomplete. It is also important to ensure that no requirements have been introduced into the design that were not present in the original requirements. Systems design has two dotted arrows, one across to systems testing and the other back to requirements specification. The horizontal arrow defines the systems testing that needs to be carried out to validate the design. The arrow to the requirements specification indicates that the design process may discover errors in the systems requirements. For example, gaps may be found in the definition of the requirements or one requirement may be found to be inconsistent with another. The same processes are applied to each stage of development.

## 5.7.2 Process requirements

The importance of quality criteria for each product was stressed earlier. Within a project, as we saw in Chapter 2, a product is the output from a process or activity. The qualities that are required in a product are captured by following the correct processes that create the product. This makes it necessary to specify for each stage, and within each stage for each activity, the **process requirements**:

- **Entry requirements** state what must exist before the stage or activity can begin. For example, before design can begin, the requirements documentation must be agreed and signed off and any design techniques to be used must be specified. If new techniques are being introduced, then the necessary training should have been given. If design begins before requirements are agreed, this may lead to problems later on if the requirements need to be changed when the design is already under way. Either the design will have to be reworked, with additional cost, or the need to amend the design may be forgotten and only picked up much later, this time with serious additional cost.

- **Implementation requirements** define how the process should be done. For example, in design, implementation requirements may specify the use of techniques such as entity modelling or logical process modelling. It also specifies the software tools that are to be used.

- **Exit requirements** indicate what should be in place for a successful sign-off of the activity. The exit requirements for design may include:
  - the documentation is complete;
  - the documentation has been reviewed;
  - the documentation meets the standards agreed;
  - the documentation covers all the requirements for this component;
  - there are no outstanding issues.

### 5.7.3   Defect removal process

Before removing a defect, it has to be identified. This is relatively straightforward, if potentially more costly, in the later stages of development when test cases can be run though the system to see if they give the expected results (**dynamic testing**). In the earlier stages of a project, different techniques have to be applied:

- desk checking;
- document review;
- peer review;
- inspection;
- walkthrough;
- static testing.

When specifying quality criteria that apply to, say, a software specification the technique for determining if those criteria have been met should be specified along with the type of staff who will implement the technique and the resources they will need.

### Desk checking

**Desk checking** is the most basic of the review activities. Authors, or creators, of products review what they have produced. This may mean reading it through checking for errors, in the case of a document, or working through the logic of a program to identify mistakes. It will, hopefully, remove many of the trivial errors before the product is subjected to more vigorous scrutiny.

**Read through the following table and identify any errors:**

*Activity 5.5*

| Data | Item range | Cross–checks |
|---|---|---|
| Time:hours | 0–23 | |
| Time:minutes | 0–60 | |
| Time: seconds | 0–60 | |
| Day | 1–31 | Cannot be greater than 30 if month = 2, 4, 6, 7, 9 or 11 |
| Month | 1–12 | |
| Year | >2000 | |

### Document review

In a **document review,** one or more people read a document to ensure that it meets the specified quality criteria. For example, is it complete, unambiguous, self-consistent and consistent with related documents? Is it clear? Are all technical terms properly used and understood? Is it to the agreed layout? If the document is a requirement, then such a review could well be carried out by the potential user of the system. For some 'structured' systems development methods, there are tools available which can capture documents (usually in the form of diagrams) and ensure that they are self-consistent. This does not remove the need to review them manually as the tool will not be able to identify completely missing requirements.

### Peer review

A **peer review** is often carried out on a design document. The author's co-workers (or **peers**) examine copies of the document and make comments about it. The issues considered include:

- Is the proposed design technically feasible?
- Is there an easier or better way to achieve the same objectives?
- Does the design conform to company standards for such processes?
- Can the design communicate with other parts of the system?
- Does the design cover all the requirements that should be included in this part?
- Are there any ambiguities?

## Inspection

An **inspection** is a much more formalized way of reviewing a product. Its purpose is to review the product in order to identify defects in a planned, independent, controlled and documented manner. It is a process with a structure similar to the following:

- Preparation:
  - setting up the review meeting, including the time, place and who is to attend;
  - distributing documentation, for example, the product and its description;
  - reviewers annotate the product, if it is a document, and record defects.

- Meeting:
  - discussion of potential defects identified by reviewers which should confirm whether they are defects or not, but not seek to produce a solution;
  - agreement of follow-up appropriate to each defect.

- Recording:
  - follow-up actions and responsibilities;
  - agreement of outcome and sign-off if appropriate.

- Follow-up:
  - advising the project manager of the outcome;
  - planning remedial work;
  - signing off when complete.

There are four roles within this process:

- The **moderator** sets the agenda and controls the review process, ensures actions and required results are agreed and, once the process is complete, signs off the review.
- The **author** provides the reviewers with relevant product documentation, answers questions about the product during the review and agrees actions to resolve defects.
- **Reviewers** undertake the review, assessing the product against the quality criteria, identifying potential defects and ensuring that the nature of any defect is clearly understood by the author.
- The **scribe** takes notes of the agreed actions, who is to carry them out and who is to check corrections and confirms these details at the end of the meeting.

## Walkthrough

A **walkthrough** is a similar formal process, but this time the author takes the audience through the documents and they feed back their comments on it. The audience may be drawn from both technical and user groups. Each would have a different view on the documents. For example, IT infrastructure management staff may be concerned about the proposed system's impact on what they already manage. Again there would be a scribe to record any actions agreed.

With peer reviews, inspections and walkthroughs, no attempt should be made during the meeting to solve problems identified. Problems should be recorded and the author then goes away and seeks to come up with a solution. The review can then be repeated if it is felt that the changes required were of sufficient significance. An alternative is for one person to be instructed to ensure that the necessary alterations have been made.

## Static testing

**Static testing (dry running)** is similar to dynamic testing, except there is no machine. It can be applied to an application at any time during its development. The idea is that test data is manually worked through the system as documented to see if it produces the expected results. Developers may wish to undertake such testing before the software is passed to a third party for testing. However it is better done by someone other than the author as it is very difficult for people to see their own mistakes.

There are software tools (programs) that can help static testing by analyzing the structure of the code. Such tools look at the branches and loops in a program and any loose ends and come up with a measure of complexity. The more complex a software component, the more difficult it is likely to be to maintain. In addition, dynamic analysis can identify code which is not executed by the test data used. This is useful as it may show up a shortcoming in the test data or redundant code which has been inserted into a program.

All the above processes take place during the activities on the left hand side of the V model. The quality control processes on the right hand side are dominated by dynamic testing.

## 5.8  Dynamic testing

Dynamic testing is divided into various levels:

- unit;
- integration;
- systems;
- user acceptance;
- regression.

For each type of testing, a set of test data and a set of expected results must be produced. Referring back to the V model, a test plan should have been produced at the appropriate stage in the development process. The plan should contain the necessary guidance for producing the test data, if not the test data itself.

## Unit testing

**Unit testing** is the very basic testing to be carried out on a piece of software. Test data for unit testing should be designed to cover the usual range of input expected for the system. Each function that the software is expected to handle should be tested. The test should take place not just once but in various combinations and different sequences. Often problems lie in the combinations of data and transactions. Testing is then extended to cover, for example, numbers just inside and just outside any specified limits. Alphabetic fields are tested with entries longer than that which the system should permit. Mandatory fields are left blank. This is not an exhaustive list but is indicative of what is necessary. All tests are designed to ensure that this particular unit will not fail because of bad data or unusual combinations but that it will handle them in a predefined way.

**Assume that the table of time/date checks drawn up (and corrected) in Activity 5.5 has now been implemented in an input screen in an IT system. Draw up a set of test data and expected results that could be used to test that the checks on data are being carried out correctly.**

*Activity 5.6*

Each set of test results should be carefully checked against the expected results to ensure agreement. Any discrepancies should be noted and reported so that the software can be amended and retested. Sometimes, however, the expected results are wrong! It is important to keep records of all faults found and resulting changes made so that if problems arise at a later stage there is a trail which can be followed to establish how they were introduced.

There are tools – commonly called **test harnesses** – which can help simulate programs or subroutines on either side of the module under test and record data input, output and the routes through the program which have been exercised. Automated tools can also simulate keyboard input and capture and compare screen output with expected results.

Once unit testing has been successfully completed the unit can be signed off and registered as a configuration item (see Chapter 4).

## Integration testing

**Integration testing** involves linking a number of components of the system and running them as a whole. This is to determine whether or not the units communicate properly with each other. The sort of things that come to light at this stage are for example:

- data items which are treated as being in different formats in different units;

- codes given different meanings by different authors;

- conditions set by one component that another cannot cope with.

Some of these can be avoided by the use of databases and database management systems (DBMS) but transaction files which link different sub-systems can still be mistakenly defined to have different formats in different places.

As errors are found they will be recorded and corrected. The integration test will then need to be repeated.

## Systems testing

**Systems testing** is the final stage in testing by the development professionals. It involves running the whole system on the infrastructure that will be used when the system is operational. It may sound just like a step up from integration testing but there are many additional issues which must be addressed, for example:

● Does the system run on the infrastructure to be used for the final system?

● Are the response times within the tolerances set in the requirements specification?

● Can the system cope with the planned workloads?

● What is the effect of high loading on the system?

Again, there are tools to help in this area. They can be used to simulate large numbers of users and high volumes of data.

### User acceptance testing

**User acceptance** testing is the crucial test. Can the users operate the system? Does it meet their expectations, not just their requirements? Users should be involved in the development process from the beginning of the project and they should have had sight of how the system is working before this point is reached so it should contain no surprises for them at this stage. Expectations as well as requirements have to be managed. It may or may not be helpful to involve users in earlier testing activities: it is a matter of keeping a balance between helping then to understand the way the system works and them seeing all the faults that arise during testing and becoming disillusioned.

It is at this stage that the importance becomes apparent of specifying clearly, right at the beginning of the project, the requirements and how they are going to be measured. Users should have a well-defined acceptance test plan that can be worked through to establish that the system meets the agreed requirements. Where discrepancies are found, the fault must be recorded, the source of the problem established and the system reworked and retested.

### Regression testing

**Regression testing** is quite different from the earlier testing processes. Regression testing needs to take place at each stage of testing. Whenever a fault is found and the offending piece of software identified it has to be corrected. Unfortunately this may well introduce further errors or uncover ones that were masked by the first error. To guard against this it is necessary to do regression testing. This involves running an agreed set of test data through the system again to confirm that not only has the original error been corrected and but that no further errors have been introduced or uncovered. The later a fault is discovered in the testing hierarchy the more regression testing has to take place.

Regression testing is also necessary whenever a change is made to the system, whether that is during development or after implementation as, all too often, changes are made to one part of a system which adversely affect another. Regression testing can largely be automated by using a standard set of test data and expected results.

## 5.9  Evaluating suppliers

It is very common these days for systems to be developed by third parties. It is therefore important to be able to establish if those third party suppliers have the necessary quality procedures in place to ensure that the software to be supplied is up to the standard expected.

To do this one has to investigate the supplier, ask some searching questions, examine documentation and perhaps undertake some auditing of their processes. The types of questions to be asked include the following.

- Do they operate to ISO 9001 (the international standard for quality management systems)?

- How are their projects organized? (For example, do they follow PRINCE 2, the standard for project management procedures sponsored by the UK government?)

- Do they use a defined development methodology, such as the Dynamic System Development Method (DSDM)?

- How is quality control exercised?

- At what points are quality reviews held?

- Is there a quality assurance process?

- Is there a configuration management system in place?

- How are change requests handled?

This list is by no means exhaustive but it gives an indication of what to ask. Following the questions, it is instructive to look for supporting documentary evidence. Are there product specifications with clearly defined quality criteria? Are there review documents? Participating in some of their reviews can increase confidence in their quality processes.

It must be recognized that the supplier and the customer have different business objectives. To bring a project to a successful conclusion therefore requires that both parties should see the project as a joint venture. The customer cannot simply hand it over to the supplier and wait for delivery, nor should the supplier hide the development process from the customer. Each side needs to be involved to an appropriate level for the type of project being undertaken.

## 5.10 ISO 9001:2000

ISO 9001:2000 is the international standard for quality management. It is part of the ISO 9000 series and is specifically aimed at producers and suppliers of products and services, not necessarily software. ISO 9000 defines the terms used while ISO 9004 applies to the improvement processes.

It should be noted that ISO 9001 states that quality should be specified, but it does not say what quality is. For example when producing a ball point pen it could be stated that a quality requirement is for the ink in the pen to last three years, but equally it could state three days. If the requirement is met then the quality has been achieved even if that quality is very low. Thus having a set of ISO 9001 procedures only guarantees that a level of quality has been specified, not that this level is accepted universally as being appropriate. Only when it is applied in a sensible fashion can there be any degree of confidence in the process.

ISO 9001:2000 is based on the following principles:

- understanding and meeting or exceeding the customer requirements;

- providing leadership for the organization that will give it purpose, unity and direction to achieve quality objectives;

- involvement of staff at all levels;
- attention to individual processes which produce intermediate or deliverable products;
- attention to inter-related processes producing deliverable products;
- continuous improvement;
- factually based decision-making;
- building mutually beneficial relationships with suppliers.

The TickIT Guide is designed to show the application of ISO 9001 to software systems. Its current version, Issue 5.0, is in line with ISO 9001:2000. TickIT applies whenever software development is carried out, the software is incorporated in the delivered product or service and the organization wishes to be ISO 9001 accredited. TickIT is managed and maintained by BSI.

## Sample questions

1.  'Fitness for purpose' defines which of the following?

(a)  The quality of the project deliverables;

(b)  The usability of the delivered IT application;

(c)  The capability of the staff who will implement the IT application;

(d)  The capability of the staff who will use the system when it is operational.

2.  User acceptance testing is an example of which of the following?

(a)  Project control;

(b)  Project monitoring;

(c)  Quality control;

(d)  Quality assurance.

3.  Which of the following is NOT a defined role in peer reviews?

(a)  The moderator;

(b)  The project manager;

(c)  The scribe;

(d)  The author.

4.  ISO 9001:2000 is a standard that defines which of the following?

(a)  Project management standards;

(b)  IT project deliverables;

(c)  Quality management systems;

(d)  Software testing procedures.

# Answers to sample questions

1. (a)    2. (c)    3. (b)    4.(c)

# Pointers to activities

## Activity 5.1

A product description must contain the following headings or sections (from Chapter 2):

- identity;
- description;
- derived from;
- components;
- format;
- quality criteria.

Additional headings may include the following:

- author;
- owner;
- date first compiled;
- date of last amendment;
- version number.

These criteria meet the requirement of being specific (a list of contents), measurable (the sections are either there or not) and achievable (it is possible, without difficulty, to ensure that each heading is present).

The quality of a product description will most likely be measured through a review process, e.g. inspection by a peer. What this process does not do is to ensure that the correct information is entered for each heading. This may require a further set of quality criteria which again should be subject to review.

## Activity 5.2

| Criterion | Good/Bad | Reason |
|---|---|---|
| Screen layouts should be consistent. | Good | This is clear and measurable. |
| Screens should be user friendly. | Bad | This is subjective and therefore not measurable. |
| The system should be able to handle 50 transactions. | Bad | This is not measurable as there is no indication as to the period of time within which the transactions have to be handled. |
| The system should allow for 20 users at any one time without degradation. | Good/Bad | This is acceptable providing there is a benchmark for processing to start with. |
| The response time should not be longer than three seconds. | Good/Bad | This is a mixture. It is clear that a response time of three seconds or less is required but it does not specify under what conditions. |

Ideally the last two criteria should be combined so that a baseline of three seconds is given with a loading of 20 simultaneous users. This can still be improved upon. The response time should be less than four seconds for at least 95% of the time with a loading of 20 simultaneous users and never exceed 10 seconds. Such a statement does allow for the odd occasion when the three seconds might be exceeded. These examples show how difficult, yet important, it is to get the quality criteria correctly specified for each product in the development process.

## Activity 5.3

A system would be maintainable if it satisfied the following criteria:

● The structure of the software is clear.

● The names used for items of data and procedures are indicative of what is being referred to.

● Code is clear and unambiguous.

● Documentation is present to support code.

The measurement for these criteria would probably be a peer review process.

## Activity 5.4

The reliability of a system can be measured in two ways:

● **Meantime between failure** (MTBF) specifies how long the system runs without failing. These days, this would be specified in weeks or even months. Something which fails every day would not be very popular with those operating the system.

● **Meantime to repair** (MTTR) specifies how long it takes to repair the system when it fails. It is no good if a system cannot be restored to a working situation in a reasonable length of time. That length of time needs to be specified as part of the acceptance criteria. Note that this measure is also related to the attribute of maintainability.

## Activity 5.5

The errors include:

● Time: minutes should be in the range 0–59.

● Time: seconds should be in the range 0–59.

● Day: July (month 7) has 31 days, February never has more than 29 days and there is no leap year check. The cross-check should be 'If month = 2 and year is leap, up to 29; if month = 2 and year is not leap, up to 28; if month = 4, 6, 9 or 11, up to 30'

## Activity 5.6

| Data item | Test description | Input | Expected result |
|---|---|---|---|
| Day | Not numeric | XX | Error message |
| | Outside lower range | 0 | Error message |
| | Inside lower range | 1 | Accept |
| | Outside upper range | 32 | Error message |
| Month | Not numeric | July | Error message |
| | Outside lower range | 0 | Error message |
| | Inside lower range | 1 | Accept |
| | Outside upper range | 13 | Error message |
| | Inside upper range | 12 | Accept |
| | Cross-check with day | day = 31<br>month = 1,3,5,7,8,10,12 | Accept |
| | Cross-check with day | day = 31<br>month = 2,4,6,9,11 | Error message |
| | Cross-check with day | day = 30<br>month = 4,6,9,11 | Accept |
| | Cross-check with day | day = 30<br>month = 2 | Error message |
| Year | Not numeric | xx | Error message |
| | Outside lower range | 2000 | Error message |
| | Inside lower range | 2001 | Accept |
| | Cross-check with day | day = 29<br>month = 2<br>year divisible by 4 | Accept |
| | Cross-check with day | day = 29<br>month = 2<br>year not divisible by 4 | Error message |

You may find some 'holes' in the above test data. It should illustrate that although devising test data is not the most glamorous job, creating effective test cases does require the kind of attention to detail that we normally expect of software developers. Devising test data will also trigger questions about the precise nature of the requirements, for example, is there really no upper limit on year?

# Estimating

## LEARNING OUTCOMES

When you have completed this chapter you should be able to demonstrate an understanding of the following:

▶ the effects of over- and under-estimating;

▶ effort versus duration;

▶ the relationship between effort and cost;

▶ estimates and targets;

▶ use of expert judgement, including advantages and disadvantages;

▶ the Delphi approach;

▶ top-down estimating

▶ bottom-up estimating;

▶ the use of analogy in estimating.

## 6.1 Introduction

In Chapter 2, we explained how to draw up a plan for a project. One of the things that we did was to allocate an estimated duration to each of the activities to be carried out. This allowed us to calculate the overall duration of the project and to identify when we would need to call upon the services of individuals to carry out their tasks. In this chapter, we will explore further the way in which these estimates can be produced.

## 6.2 What we estimate and why it is important

### 6.2.1 Effort versus duration

As well as estimating the time from the start to the end of an activity, it is also necessary to assess the amount of effort needed. Duration should not be confused with effort. For example, if it takes one worker two hours to clear a car park of snow, then, all other things being equal, it takes two workers only one hour. In both cases, the **effort** is two hours but the activity **duration** is two hours in one case and only one hour in the other. There can be cases where the duration is longer: for example, where someone only works in the afternoons on a particular task. In fact, a problem is that activities often take longer than planned even though the effort has not increased. This may happen, for instance, when you have to wait for approval from a higher level of management before a job is signed off. This distinction between effort and duration can be particularly important when assessing the probable cost of a project as staff costs are often governed by the hours actually worked.

### 6.2.2    The effects of over- and under-estimating

If effort and duration are under-estimated, the project can fail because it has exceeded its budget or has been delayed beyond its agreed completion date. This may be so even when staff have worked efficiently and conscientiously. Allocating less time and money than is really needed can also affect the quality of the final project deliverables: team-members may work hard to meet deadlines but, as a consequence, produce sub-standard work.

On the other hand, estimates that are too generous can also be a problem. If the estimate is the basis for a bid to carry out some work for an external customer, then an excessively high estimate may lead to the work being lost to a competitor. Parkinson's Law (work expands to fill the time available) means that an excessively generous estimate may lead to lower productivity. If a task is allocated four weeks when it really needs only three, then there is a chance that, with the pressure removed, staff will actually take the planned time.

### 6.2.3    Estimates and targets

Identifying the exact time it will take to do something is very difficult because, if the same task is repeated a number of times, each instance of the task execution is likely to have a slightly different duration. Take going to work by car. It is unlikely that on any two days this will take exactly the same amount of time. The journey time will vary because of factors such as the weather conditions and the pressure of traffic. This means that an estimate of effort or time is really a most likely effort/time with a range of possibilities on each side of it. Within this range of times we can choose a **target** – we can go for an 'aggressive' target which may get the job done quickly, but with a strong possibility of failure, or a more generous estimate which is likely to expand the length of time needed, but with a safer chance of the target being met. The target, if at all reasonable, can become a **self-fulfilling prophecy** – with the commuting example, if you know that you are going to be late you may take steps to speed up, perhaps by taking an alternative route if the normal one is congested.

## 6.3    Expert judgement

### 6.3.1    Using expert judgement

Where do you start if you want to produce reasonable estimates? Although estimating is treated as a separate, isolated, topic in project management and information systems development, it in fact depends on the completion of other tasks that provide information for estimates. For a start, you need to know:

● What activities are going to be carried out during the course of the project?

● How much work is going to have to be carried out by these activities?

For example, to work out how long it will take to install some software on all the work-stations in an organization, we need to know approximately how long it takes to install the software on a single work-station and how many work-stations there are in the organization. We may also need to know how geographically dispersed the work-stations are. The best person to tell us about these things would be someone familiar with the tasks to be carried out and the environment in which these tasks are to be carried out. As a general rule, the best people to estimate effort are those who are experts in the area. As a consequence, most guides to estimating identify **expert opinion** or **expert judgement** as an estimating method.

Although 'phoning a friend' can be a very sensible approach, there remains the question of how the experts themselves derive their estimates. There is a possibility that they have their own experts upon whom they can call, but at some point someone must sit down and work out the estimate based on their own judgement.

The advantages of using expert judgement include the following:

- It involves the people with the best experience of similar work in the past and the best knowledge of the work environment.

- The people who are most likely to be doing the work are involved with the estimating process – they will be more motivated to meet the targets set if they have had a hand in setting them in the first place.

There are, however, some balancing risks:

- The task to be carried out may be a new one of which there is no prior experience.

- Experts can be prone to human error – they may, for example, underestimate the time that they would need to carry out a task in case a larger figure suggests that they are less capable.

- It can be difficult for the project planner to evaluate the quality of an estimate that is essentially someone else's guess.

- Large complex tasks may require the expertise of several different specialists.

### 6.3.2   The Delphi approach

One method that attempts to improve the quality of expert judgement is the **Delphi technique** which originated in the Rand corporation in the USA. There are different versions of this, but the general principle is that a group of experts are asked to produce, individually and without consulting others, an estimate supported by some kind of rationale. These are all forwarded to a moderator who collates the replies and circulates them back to the group as a whole. Each member of the group can now read the anonymous estimates and supporting rationales of the other group members. They may now submit a revised opinion. Hopefully, the opinions of the experts should converge on a consensus.

The justification for the technique is that it should lead to people's views being judged on their merits and that undue deference will not be paid to more senior staff or the more dominant personalities.

## 6.4   Approaches to estimating

We are now going to discuss key approaches to estimating. However first we are going to explain the terms **bottom–up** and **top–down**. Note that these are not specific estimating methods, but describe a way of grouping estimating methods.

### 6.4.1    Bottom-up

In essence, we break the task for which an estimate is to be produced into component sub-tasks and then break the component sub-tasks into sub-sub-tasks and so on until we get to elements that we think would not take one or two people more than a week to complete. The idea is that you can realistically imagine what can be accomplished in one or two weeks in a way that would not be possible for one or two months. To get an overall estimate of the effort needed for the project, you simply add up all the effort for the component tasks.

This method is also sometimes called **analytical** or **activity-based estimating**. Some people find the name 'bottom-up' confusing (especially those who are or who have been software developers) because the first part of the process is really top-down!

**Which planning product identified in Chapter 2 could be the basis for an initial bottom-up estimate?**          *Activity 6.1*

A bottom-up estimate is recommended where you have no accurate historical records of relevant past projects to guide you. A disadvantage of the method is that it is very time-consuming as you have, in effect, to draw up a detailed plan of how the project is to be carried out first. It could be argued that you are going to have to do this anyway. However, this may be very tedious and speculative if you have been asked for a rough estimate at the feasibility study stage of the project proposal.

**You have been asked to organize the design and implementation of training for the Canal Dreams booking system. Identify the component activities in this overall task, as you would for the first stage of the bottom-up approach to estimating effort.**          *Activity 6.2*

### 6.4.2    Top-down

With the top-down approach, we look for some overall characteristics of the job to be done and from these produce a global effort estimate. Nearly always this figure is based on our knowledge of past cases.

An example of top-down estimating is when a house-owner has to make a decision about the sum for which they should insure their house. The question here is the probable cost of rebuilding the house in the event of it being destroyed, for example by fire. Most insurance companies produce a handy set of tables where you can look up such variables as the number of storeys your house has, the number of bedrooms, the area of floor-space, the material out of which it has been constructed and the region in which it is located. For each combination of these characteristics will be a suggested rebuilding cost. The insurance company can produce such tables as it has access to many historical records of the actual cost of building houses.

This is essentially a top-down approach because only one global figure is produced. In the unhappy case of fire actually occurring, this figure would not help a builder to calculate how much effort would be needed to dig the foundations, build the walls, put on the roof and all the other individual components of the building operation. A builder may be able to use past experience of the proportion of total costs usually consumed by each type of activity, such as foundation digging.

## 6.5 A parametric approach

The base estimate created when using a top-down approach can be derived in a number of ways. In estimating the costs of rebuilding a house, a **parametric** method was used. This means that the estimate was based on certain variables or parameters (for example, the number of storeys in the house and the number of bedrooms). These parameters can be said to 'drive' the size of the house to be built: you would expect a house with three storeys and five bedrooms to be physically bigger than a bungalow with only two bedrooms. These parameters are therefore sometimes called **size drivers**. You would also expect the three-storey building to need more work, or effort, to build than the bungalow. These parameters are also sometimes called **effort drivers**.

### 6.5.1 Size drivers and productivity

Earlier we had an example where a technician was allocated the job of installing a new piece of software on every work-station in an organization. Clearly the more work-stations there are, the bigger the job and the longer its duration. Hence the number of work-stations is a size driver and an effort driver for this activity.

Identify the possible size and effort drivers in the Canal Dreams booking system for each of the following activities:

*Activity 6.3*

(a) Creating training material for users;

(b) Training users;

(c) Carrying out acceptance tests;

(d) Setting up standing details such as boat records.

In order to produce an estimate of effort using this method we also need a productivity rate. For example, in addition to the number of work-stations we would need to know the average time needed to install the software on a single work-station. This time per work-station would be the **productivity rate.** If this rate was 12 minutes per work-station and there were 50 work-stations then we could guess the overall duration of the job would be around 50 × 12 minutes, that is about ten hours.

The usual way to obtain the productivity rate is from records of past projects. Where these are not available within an organization, it is sometimes possible to obtain 'industry' data that relates, not to projects in a single organization, but to projects in a particular industrial sector. This kind of information can help managers compare the productivity in their organization with that of others – this is sometimes called **benchmarking.** If they find that they have much lower productivity, this may spur them on to search for more productive ways of working. However, caution needs to be practised if the reason for using industry data is because local project data is missing: there can be large differences in productivity between organizations, because organizations and their businesses are so different.

*Activity 6.4*

**In the earlier example about the time needed to drive to work, identify:**

**(a) the size driver;**

**(b) the productivity rate;**

**(c) other factors that may cause a variation in the time it takes to get to work.**

The additional factors may be called **productivity drivers.** A key productivity driver when it comes to developing and implementing IT systems is experience. When putting a figure on how long a technical activity like developing software code is going to take, more experienced estimators will try to find out who will be doing the work.

Productivity drivers vary from activity to activity, but other drivers often include:

● the availability of tools to assist in the work;

● communication overheads, including the time it takes to get requirements clarified and approved;

● the stability of the environment, i.e., the extent to which the work has to cope with changes to requirements or resources;

● the size of the project team: there is a tendency for larger jobs involving lots of people to be less efficient than smaller ones because more time has to be spent on management, planning and co-ordination at the expense of 'real work'.

The problems that can affect productivity are often considered at the same time as risks to the project in general (see Chapter 7).

### 6.5.2   Function points

There was a time when almost all IT projects involved writing software of some description. This is now increasingly less so for many reasons, one of which is the tendency to use 'off-the-shelf' software applications. However there are still many cases where software has to be written specially and these situations can cause particular challenges for the estimation of development effort.

If we use a parametric approach, the first question is what to use as size drivers. If IT is old enough to have any real 'traditions', then one of the longest-established of these would be to use **lines of code** as the size driver for software development. (When software is written, the programmer writes the instructions – as lines of code – in a form which is comprehensible to human experts. This 'code' is an electronic document which can be changed, added to and printed. When the code is to be executed by the computer, the document is 'read' by a special piece of software which converts it into a format that the computer can interpret automatically.)

From this very brief explanation it can be seen that:

● the code is a very technical product – it would need a software expert to estimate the number of lines of code;

● you will not know the exact number of lines of code until quite near the end of the project; most other size drivers are known at the beginning, or at least at an early stage, of the project.

Things are also complicated by there being many programming languages. Some are more 'powerful' than others, that is, they need fewer lines of code to carry out a particular procedure.

Rather than use this technical unit of size which is invisible to everyone except the software developers, it is more convenient to use as the size drivers counts of features of the software application that are externally apparent. This would be rather like using the number of storeys, the floor space and so on to estimate the cost of a house rather than the number of bricks. With software applications, this can be done with **function points**.

For the purposes of the Foundation Certificate, you do **not** need to know the details of the rules of function point counting. There are at least two major systems of function point counting and some of the detailed rules are rather arcane, to say the least. The following description should be enough to give a general idea of the approach. (It is based on one particular version – Mark II, or Symons, function points – simply because this is, in our view, the simplest method for getting an understanding of the general principles of the approach.)

(a)    The size drivers are features of the software that are apparent to the user. In general, users are aware of the **transactions** that they can carry out when using a software application. A transaction is where the user **inputs** something into the computer (normally by typing), the computer carries out a procedure and comes up with a **response**, normally in the form of an **output,** and the computer system is left in a **consistent state.**

When booking a boating holiday using the new Canal Dreams booking system, you make a mistake (for example, typing in an invalid date) and an error message is displayed. Although an input has been followed by an output (the error message), the system is not in a consistent state: only half the booking details have been set up. In this case the processing so far would not be regarded as a transaction: either the whole of the booking would be rejected or the processing would continue until a complete and correct set of booking details had been input.

(b)    For each transaction, a count is made of the number of items of information that are input and output, and the number of tables of information that are accessed. In general, it can be assumed that the more of these there are, the more lines of code will have to written, and the more work there will be for the system developers.

(c)    The counts are weighted to take account of the relative difficulty of implementing each type of feature. For example, a simple output is normally easier to implement than an input. With inputs you have often to carry out error checking which adds to the programmer's work. To take account of these differences in difficulty, the feature counts are weighted appropriately. In the Mark II method, inputs are weighted 0.58, outputs 0.26 and entity accesses 1.66. Effectively this means that the weighting between inputs, outputs and entity accesses is about 2:1:6. The peculiar numbers used are because the inventor of this method wanted the resulting function point counts to be about the same as for the American method (specified by the international function point user group, IFPUG) and hoped to achieve this by making the weightings add up to 2.50 (that is 0.58 + 0.26 + 1.66).

### 6.5.3    An example of function point counting

Within the Canal Dreams booking system, there is a transaction which records a payment made by a customer. There are **three inputs** for a new payment:

● date;

● customer account reference;

● amount.

There are **four** possible **outputs** from the transaction:

● payment reference, a number allocated automatically by the computer system;

● customer name;

● customer address;

● an error message.

To carry out this transaction, a CUSTOMER ACCOUNT table and a PAYMENT table are accessed, giving **two** entity accesses. The function point count for this transaction is therefore:

$$(3 \times 0.58) + (4 \times 0.26) + (2 \times 1.66) \text{ i.e. } 6.10.$$

What does this 6.10 really represent? It is best regarded as an index value that gives an idea of the amount of processing that is carried out by the transaction. For a single, isolated, transaction, this measure is not very accurate. However, if you were able to add up the function point counts for all the transactions in an information system, then it is likely that the count for the application as a whole would be a useful indicator of its size.

We can use a function point count to find out the relative productivity of development projects that have already been completed. We may find that the average number of function points implemented per day is around five. This may not seem a huge number, but 'development effort' here includes the whole development cycle, from requirements gathering to testing. When a new project proposal comes along, a preliminary investigation may suggest that the delivered system would have a count of about 250 function points. The estimated effort may therefore be in the region of 250/5 days, that is 50 days.

## 6.6    Estimating by analogy

The function point approach (and, indeed, the more generic approach of size driver and productivity rate) is based on the assumption that there exists a structured record of the details of past projects. Often, however, such a record does not exist. For smaller organizations, particularly, the IT projects that have been previously implemented may all seem to have their own peculiarities. For example, some may have involved the installation of off-the-shelf packages; others may have required specially written software, some a mixture of the two, and so on. This may seem to imply that previous experience is not a stable basis for estimating the effort for new projects. However, in this kind of situation the **analogy** or **comparative** approach could be used.

The main steps with this method are as follows.

(a)    Identify the key characteristics of the new project.

(b)    Search for a previous project which has similar characteristics.

(c) Use the actual effort recorded for the previous project as the base estimate for the new one.

(d) Identify the key differences between the old and the new projects (it is unlikely that the old project is an exact match for the new one).

(e) Adjust the base estimate to take account of the identified differences.

An analogy approach can be used to create a top-down estimate for a project. Where there is no past project which seems to be a useful analogy for the new project, an estimator may use analogy to select parts of old projects that seem similar to components of the current project (using analogy as part of a bottom-up approach).

As Table 6.1 shows, both analogy and the parametric approaches can be used either at the overall level of a project or for estimating the effort needed for components. The activity-based approach – breaking down the overall task into smaller components – seems almost by definition to be a bottom-up approach.

**Table 6.1 Relationship between top-down/bottom-up and the three main estimating approaches**

|  | activity–based | analytical parametric | analogy comparative |
|---|---|---|---|
| top down |  | ✓ | ✓ |
| bottom-up | ✓ | ✓ | ✓ |

## 6.7 Checklist

As a project planner you may often need to use the effort estimates produced by experts from technical areas where you are not knowledgeable. Are there any ways in which you can realistically review these estimates? It may be possible to assess the plausibility of the estimates by asking the estimator the questions below.

● What methods were used to produce the estimates?

● How is the relative size of the job measured (in other words, what are the size/effort drivers)?

● How much effort was assumed would be required for each unit of the size driver (in other words, what productivity rates are you assuming)?

● Can a past project of about the same size be identified which had about the same effort?

● If a job with a comparable size cannot be identified, can past jobs which had similar productivity rates be found?

## Sample questions

1. An under-estimate of effort is MOST likely to lead to which of the following?

(a) decreased productivity;

(b) decreased quality;

(c) a less competitive bid for a contract;

(d) a longer project duration.

2.  Which of the following estimating methods is MOST likely to be used bottom-up?

(a)   parametric;

(b)   algorithmic;

(c)   Delphi;

(d)   activity-based.

## XYZ organization Scenario

Staff have managed to develop information systems at a rate of five function points per staff-day. A new system has been assessed as requiring 120 function points to implement, but the staff available are relatively inexperienced and are only 80% as productive as the staff usually used in such projects.

**Both questions 3 and 4 use this scenario.**

3.  In the XYZ scenario, of which of the following is 80% the value?

(a)   a size driver

(b)   an effort driver

(c)   a productivity rate

(d)   a productivity driver

4. In the XYZ scenario, what would be the best estimate of effort for the project?

(a) 30 days

(b) 25 days

(c) 24 days

(d) 20 days.

## Answers to sample questions

1. (b)    2. (b)    3. (d)    4. (a)

## Pointers to activities

### Activity 6.1

The work breakdown structure (or possibly the product breakdown structure).

### Activity 6.2

Among the activities that may need to be carried out are:

● Analyze package to document the main user functions.

● Collect information about training requirements from users/user management.

- Draw up a course plan.
- Design presentations.
- Design practical exercises.
- Write training handouts.
- Arrange dates for course delivery.
- Book accommodation and equipment.
- Notify attendees.

## Activity 6.3

The following are suitable answers:

(a) The number of functions that users need to be able to use.

(b) The number of functions that users need to be able to use and the number of users to be trained.

(c) The number of functions to be tested and the number of input and output data items to be tested.

(d) The number of boat records to be entered and the number of items of information to be input for each record.

## Activity 6.4

(a) The size driver would be the distance driven to work.

(b) The productivity rate would be the average speed of the car.

(c) We have already suggested that the weather and the amount of traffic congestion could have an effect on the travel time.

In this case, the weather and traffic do not increase the size of the job to be done – the distance to work remains the same. These factors are best seen as influences on the productivity rate. In order to assess more accurately the time it takes me to go to work, I could take account of these intermittent constraints on my speed. I may be aware, for instance, that the rush-hour traffic in the morning tends to be significantly less during school holidays. I could therefore perhaps allow myself to start off to work a few minutes later when it is half-term. On the other hand, I may start earlier if the weather is foggy, as I know that this can slow down the traffic.

# Risk

## LEARNING OUTCOMES

When you have completed this chapter you should be able to demonstrate an understanding of the following:

▶ identification and prioritization of risks;

▶ assessment of risk exposure;

▶ risk reduction activities versus contingency actions;

▶ typical risks associated with information systems;

▶ assessment of value of risk reduction activities;

▶ maintenance of risk registers;

▶ risk reduction and the selection of appropriate project approaches such as prototyping.

## 7.1 Introduction

There are a number of definitions that are used to define risk, some embracing opportunities as well as threats. Arguably, the most used is the definition provided in PRINCE 2, the project management standard sponsored by the UK government:

**The chance of exposure to the adverse consequences of future events.**

Those adverse effects could be a reduction in the efficiency of the project, or even the failure of the project. They include higher development costs, delayed project completion, reduced scope and reduced performance. The adverse effect could also be a less effective completed system that fails to deliver the appropriate capability, which in turn means that the original business case is not fully realized.

Some risks are **project–related**, that is, they threaten the successful achievement of the project's objectives. Other risks are **business risks**. The project team may meet the project objectives and deliver the required IT application on time and within budget. Business conditions, however, may mean the users cannot use the application to achieve their business objectives. In the Canal Dreams scenario, a slump in the demand for canal holidays may mean that the investment in the new booking system does not pay for itself. It should be remembered, on the other hand, that risks involve opportunities as well as threats. The opportunity perspective will be illustrated later in the chapter.

The risks may originate from inside the development programme or project, or they could be caused by external events. It is worth noting at this point that if an adverse event has already occurred, this is not a risk, but should be treated as a **project issue,** that is, a problem that needs to be resolved.

External (or environmental) risks include:

- government intervention;
- reduction in the supply of resources, including staff;
- reduction in financial support;
- increased competition from rivals;
- social developments.

Internal risks include:

- staff changes;
- lack of policies which can guide decision-making;
- increased scope of changes;
- lack of developer experience;
- sabotage.

The objectives of risk management are to identify, address and minimize risks before they become threats to the successful completion of a project. However, we need to be aware of 'the law of diminishing returns' which suggests that the initial effort and expenditure provides the best return and that the benefits from further spending to solve a problem gradually become smaller. Buying one smoke detector for your house when you have none could make a big difference to your safety, but buying another when you already have five will probably make little difference. Actions to eliminate some risks may be prohibitively costly, difficult or lengthy and, if adopted, would adversely affect the business case.

## 7.2    Risk management

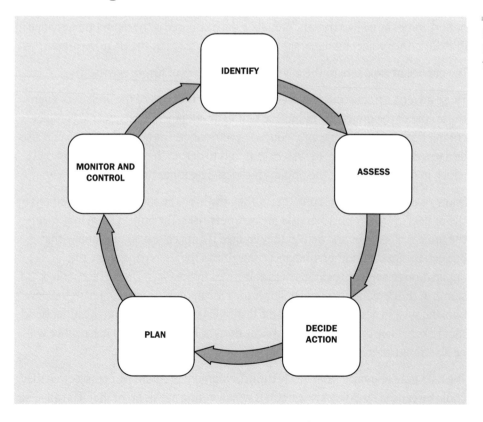

Fig. 7.1
Risk management
framework

The management of risks is similar to the management of any activity. There is a planning and control cycle similar to the one described in Chapter 3. Risk management is a continuous process throughout the life of the project. However, before managers form any plans, they need to identify the risks and make some decisions about them. A typical risk management framework is shown in Figure 7.1. Major risks have to be identified and then plans have to be made to deal with them. These plans could include activities that enable other activities to take place in the event of a risk turning into a real problem. For example, taking back-ups of important computer files allows them to be restored if the originals are damaged. The project is then executed and is monitored and controlled to see where risks have materialized and where appropriate actions need to be initiated.

## 7.3 Identifying risks

Before the risk identification process begins, there will be a number of facts or issues that are already known. These include particular views, trends, or constraints within the project development environment. It is worth examining these facts as they often provide the causes of the risks that will eventually need to be managed.

**Reread the Canal Dreams project scenario in Chapter 1 and identify any particular features of the project or its environment which you think could lead to difficulties.**

*Activity 7.1*

There are a number of aids to building the initial list of risks. Many risk text books, application development documents and company standards provide **prompt lists**, containing a number of **generic** business and project risks that originate from outside and inside the project. These can be used to determine which risks in the list apply to the project.

In addition to the generic risks that may apply to almost any project, there are **specific risks** peculiar to the circumstances of the particular project that need to be identified – most of the risks to the Canal Dreams project that we identified were fairly specific to that project. Experts or stakeholders within the project can be interviewed to elicit their ideas on potential specific risks. It can be useful to hold a brainstorming workshop with a number of stakeholders to identify risks; the discussion of a risk by one member may lead to another member recognizing a further risk. Past project documentation can be searched to provide further risks for consideration. It is important however, in view of the law of diminishing returns, that the project does not just assume all the generic risks. Instead, be discriminating and dismiss those risks that do not really belong to the project. For example, an aircraft crashing into the software application development laboratory is a risk, but not one that necessarily must be considered, unless it is under a busy flight takeoff path. Do not dilute the risk management approach by paying too much attention to every conceivable risk.

It is helpful in identifying a risk to recognize the true focus of the problem. For example, you should not say:

Development of the application software may overrun.

You should say:

There is a risk that the application programmers are too inexperienced in the chosen language and therefore the application software may be completed late.

Note that this risk statement is only appropriate when it is set in the context of the given project. In this example, it is a fact that the programmers are inexperienced in the chosen language. This fact, or issue, is not a risk but a cause of a potential risk. A risk has an element of uncertainty about it. The fact that there is limited experience of the chosen programming language only becomes a risk in our project if the difficulty of the design and coding of the application software makes demands that are too great for the inexperienced programmers.

Given the list of possible issues listed in Activity 7.1, identify six possible risks for the Canal Dreams booking system project.          *Activity 7.2*

## 7.4 Assessing the risk

### 7.4.1 Risk evaluation criteria

We recognized that we may not be able to take action for all possible risks identified. Therefore there is a need to prioritize the risks, but before this can happen there is a need to evaluate the risk itself. The evaluation is based on three criteria:

● the **probability** that the risk will occur;

● the **impact** that the risk will have, should it occur;

● the **proximity** of the risk.

### 7.4.2 Risk probability

The **probability**, or likelihood, that a risk will occur is somewhere between never happening and certain to occur. This assessment can be either **quantitative** or **qualitative.** There is a wide range of approaches to obtaining qualitative assessments including interviewing experts or stakeholders and brainstorming in a workshop. Various qualitative descriptions of probability can be used to describe the probability of a risk occurring, such as 'extremely likely', 'very high', 'high', 'medium', 'low', 'very low', 'improbable'. Similar descriptions are used in the qualitative assessment of the impact the risk will have on cost, quality or time.

However, to aid the prioritizing and decision-making associated with risk management, it is better to assign quantitative values to the probability and the impact (see Tables 7.1 to 7.4).

**Table 7.1** Mapping qualitative and quantitative assessments of risk probability

| Index | Impact level | Comment |
| --- | --- | --- |
| 4 | High | Greater than 50% chance that the risk will occur |
| 3 | Significant | 30–50% chance that the risk will occur |
| 2 | Moderate | 10–29% chance that the risk will occur |
| 1 | Low | Less than 10% chance that the risk will occur |

| Table 7.2 | Mapping qualitative and quantitative assessments of impact on cost | |
|---|---|---|
| **Index** | **Impact level** | **Comment** |
| 4 | High | Greater than 20% above the project cost tolerance |
| 3 | Significant | Up to 20% above the project cost tolerance |
| 2 | Moderate | Greater than 50% of the project cost tolerance but still within it |
| 1 | Low | Within 50% of cost tolerance |

| Table 7.3 | Mapping qualitative and quantitative assessments of impact on scope | |
|---|---|---|
| **Index** | **Impact level** | **Comment** |
| 4 | High | Inability to meet mandatory project functionality |
| 3 | Significant | Shortfalls in key functionality |
| 2 | Moderate | Shortfalls in secondary functionality |
| 1 | Low | Some minor functions missing |

| Table 7.4 | Mapping qualitative and quantitative assessments of impact on time | |
|---|---|---|
| **Index** | **Impact level** | **Comment** |
| 4 | High | Greater than 20% above project time tolerance |
| 3 | Significant | Up to 20% above the project time tolerance |
| 2 | Moderate | Greater than 50% of the project time tolerance but still within it |
| 1 | Low | Within 50% of project time tolerance |

The Delphi method (see Chapter 6) is an extended version of the expert approach to risk assessment. Risk assessment is very closely associated with effort estimation and in some cases can be carried out at the same time.

There are risk estimation techniques that can be used to provide more precise quantitative measurements. However the ISEB foundation qualification does not expect such detailed coverage and so this text will consider only qualitative values of the type used in Tables 7.1 to 7.4.

### 7.4.3 Risk impact

The **impact**, or severity, is the adverse effect that the risk will have on the project, should it materialize. This impact may be a longer development **time**, a reduction in the **scope** of the deliverable, a reduction in the **performance of the deliverable**, or an increase in the **resources needed.** The scope and the performance are often combined as a reduction in **quality**. The increased resources, both of materials and labour, are usually referred to as increased **costs**. Should a risk occur, it may impact time, cost or quality, and of course any impact will affect the business case.

It will be clear now that any phrase relating only to time, quality or cost is not the risk but the impact of the risk. For example 'the development of application software may overrun' is clearly an impact of a risk on time, rather than the risk itself. One must be clearer in identifying the source of risk, which in this example is 'the application programmers are too inexperienced in the chosen language'.

That said, it is now clearer how a risk can be viewed as an **opportunity.** Let us examine the example above, of developing application software in a chosen language. Our programmers may be very experienced in the C++ programming language, with limited knowledge of Java. The project however calls for programmers experienced in Java. The organization could recruit new programmers or contract the development out to a third party. Both those options for reducing risk are very expensive. Consequently, there is an opportunity to save some development costs by 'taking a risk' and utilizing the limited Java experience of the current programmers. Furthermore, the plan for the software development tasks may increase the expected duration of the tasks to take account of the developers' lack of experience. If however the developers were able to pick up Java very quickly their tasks may be completed earlier than scheduled. In this case the project manager ought to exploit this opportunity and start the next task as soon as possible. The time gained here will be a useful buffer if problems occur later in the project.

### 7.4.4    Risk proximity

The **proximity** relates to the time period in the project schedule when the risk may occur. The risk evaluation may identify that a given risk is more likely to occur at a certain time, that it will not happen after a certain project milestone or even that the impact could change depending on when the risk occurs. The risk of inexperienced Java programmers delaying completion of work will mainly have an impact on the software development stage. In Chapter 1, it was noted that uncertainty about a project was greatest at the beginning because of all the unknowns associated with a new project. As knowledge is gained about the application and technical domains during the project much of this uncertainty is reduced.

### 7.4.5    Prioritizing risks

Once the initial evaluation has taken place, the risks can be prioritized, to ensure that the risk management effort is placed where it is needed most. There are two aspects to consider: the ranking of risks among themselves and the overall 'riskiness' of the project. To obtain a ranking of the identified risks, some mathematical expression could be used to combine the probabilities and the impacts. In the simplest approach, a weighting is applied both to the potential damage and to the probability of loss, and the two values are then multiplied. To aid decision making for qualitative assessment, a **probability impact grid,** sometimes known as a **summary risk profile,** can be used (see Figure 7.2). In the grid the numbers uniquely identify each risk. Some organizations would have three probability impact grids to cover the impacts on time, cost and quality respectively. Other organizations combine them. Organizations refine their understanding of the risk profiles over time and can judge more accurately the threat of a risk and the action to be taken to deal with it.

The overall 'riskiness' of the project is sometimes known as the project's **risk exposure.** A monetary value can be placed on each risk, determined by the impact should the risk occur but multiplied by the probability that it will occur. For example, if the damage caused by a risk was £1,000 and there was a 10% chance of the risk occurring then the risk exposure for that risk would be £100. This can be seen as the notional amount that should be paid into a central 'insurance' fund for the project. By adding up all the values for individual risks, a total value for all the risks is arrived

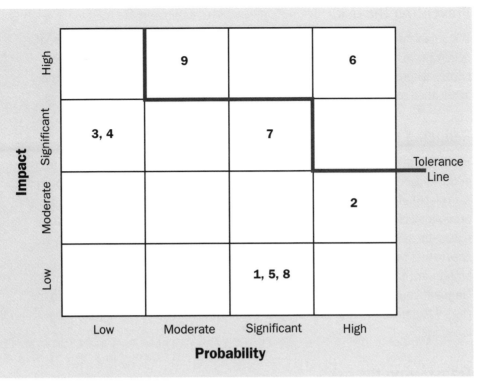

Fig. 7.2
Probability impact
grid

at for the project. This figure reflects the amount that the project should set aside to deal with possible risks emerging during the project. If an organization deems this value to be unacceptable, then the project is abandoned or actions are taken on one or more of the risks to prevent or reduce the impacts. In addition to this figure the organization may also wish to consider the 'worst case scenario' where all the risks occur.

## 7.5 Deciding the appropriate actions

The uncertainty on the project due to the risks identified can be reduced by taking **mitigating action**. Mitigation is defined as an action to reduce, transfer or eliminate a risk. Any mitigation action chosen will consequently mean revisiting the project schedule, the development costs, the functional scope, and the performance of the deliverables and updating them, if necessary, to take account of the chosen action. In addition to the option of adopting one or more mitigating actions for a given risk, the project team could decide to take no action at all or to develop a contingency plan.

### 7.5.1 Accepting the risk

The organization could accept the risk and take no further action other than to monitor it. It could be argued that taking no action is not, strictly-speaking, a form of mitigation, but it is mentioned here for the sake of completeness. This option could be chosen if it is deemed that the risk probability is low, the impact is acceptable, or it is thought that none of the actions listed below would be appropriate. The actions may not be thought appropriate due to the cost of action outweighing the impact cost or to the lack of any plausible alternative. In the case of the inexperience of programmers in Java, the cost of action was judged to exceed the cost of the impact of the risk.

### 7.5.2    Preventing the risk

The organization could take action, before the risk is due, to prevent the risk from occurring. In the example above, where the programmers' inexperience of the chosen language is a risk, the risk could be prevented by deciding to develop the application software in the language with which they are familiar.

### 7.5.3    Reducing the risk

The organization could take action to reduce the probability that the risk will occur or the impact that the risk will have if it does occur. Again any reduction action will take place before the expected risk occurs. For example, the risk of inexperienced programmers can be reduced by providing further training. If it was decided to reduce the possible impact on the project deadline, steps could be taken to try to ensure that the development was not late despite the inexperience of the programmers. For example, one or more specialists experienced in the chosen language could be recruited to join the development team to act as advisers in the case of technical misunderstandings of the features of the 'new' programming language.

### 7.5.4    Transferring the risk

The organization may take action to transfer the risk to another party. For example, the organization with inexperienced programmers could contract the work out to a software house. If the software house was working to a fixed-price contract then the problem of possible cost over-runs would be transferred to them.

### 7.5.5    Contingency

The organization may decide not to take any action before the risk occurs, but instead to plan an action to be taken once the risk occurs or if it becomes more certain that the risk will occur. A contingency plan for the project with inexperienced programmers could be to purchase application software from a vendor if it becomes clear that the delivery of the in-house software will not be on time. The major difference between this action and the other actions is that they incur costs even though the risk may not materialize. It should be noted that with all actions, there will be the costs of managing the risk (see Section 7.6). There may also be costs associated with creating the conditions which would allow the contingency action to take place, as when back-ups of files have to be taken to allow the contingency action of restoring files if they are corrupted.

**Explain the actions that may be taken to mitigate the risks to the Canal Dreams project scenario that were noted as a result of Activity 7.2.**    *Activity 7.3*

When selecting the mitigating actions to be taken, it may well be that more than one action is appropriate for a particular risk, or that a chosen action can mitigate more than one risk. For example, in the case of the Canal Dreams booking system, extensive consultation with the users at all the boatyards may not only ensure that all requirements are considered but may also reduce local staff resistance to the new system.

Because the action chosen will have an impact on the original project plans, it is important to ensure that the benefits of the action outweigh the benefits of inaction. How many actions to approve, and to which risks, is a key decision process in risk management. Initial focus is likely to be on the **show–stoppers**, risks that would prevent completion of the project. A number of plans could be adopted that would reduce the **risk exposure** of the project to a level acceptable. Alternatively actions could be taken to address just the highest priority risks. For example, a **risk tolerance line** has been shown on the probability impact grid in Figure 7.2. The organization will not approve a project with risks that occur above this line. Therefore action would be taken to ensure a new position on the grid for these risks by reducing their probability or impact or both.

## 7.6  Planning, monitoring and control

In the paragraphs above, it has been assumed that the risk identification, risk assessment and mitigating actions have occurred in the earlier stages of a project, during the project planning. However, all of the processes described above will continue throughout the life of the project, as new risks may be identified at any time and there may also be secondary risks that result from actions to reduce initial risks. For example, outsourcing software development to a software house because of the inexperience of in-house developers will itself generate new risks.

The monitoring of risks may be part of the project control cycle (see Chapter 3). The monitoring process is most likely to be a mixture of regular timed reviews and reviews held after significant events, particularly the end of a project stage

It is important to have a structured project risk plan to document the planning and to facilitate the monitoring and control process. This will consist of a **risk register,** also known as a **risk log,** which lists all the risks identified with the project (see Figure 7.3).

| Risk register | | | | | | | |
|---|---|---|---|---|---|---|---|
| Risk Id | Risk title | Risk owner | Post risk management | | | | Plan No |
| | | | Prob. | Cost | Quality | Time | |
| | | | | | | | |
| | | | | | | | |
| | | | | | | | |
| | | | | | | | |
| | | | | | | | |
| | | | | | | | |
| | | | | | | | |

Fig. 7.3
Risk register

Typical headings are shown in the form, but others could be added, for example, a risk category or the earliest date that the risk could occur. Initially, not all the entries would be completed. As risk assessment is an iterative process, entries such as the post risk management values are entered after the appropriate actions have been approved and the risk plans formulated.

For each risk in the risk register, an individual **risk record** will be raised (see Figure 7.4). Note that Figure 7.4 shows the probability and impacts both before and after any agreed mitigating action is taken – see the lines for 'pre-risk management' and

| Risk Record | | | | |
|---|---|---|---|---|
| Risk Id | Risk title | | Risk category | Plan No. |
| Risk owner | Initiation date | Earliest Occurrence | Review date | Deletion date |
| Risk description | | | | |
| Impact description | | | | |

| Probability/impact values | Probability | Impact on | | |
|---|---|---|---|---|
| | | Cost | Quality | Time |
| Pre risk management | | | | |
| Post risk management | | | | |

History

| Version | Date | Action | Comments |
|---|---|---|---|
| | | | |
| | | | |
| | | | |
| | | | |
| | | | |

Fig. 7.4
Risk record

'post-risk management'. In addition to the risk register and risk records, plans of the actions chosen need to be documented. As noted earlier there is not necessarily a one-to-one relationship between a risk and a risk plan. An individual risk may necessitate a number of plans before the risk exposure is reduced to an acceptable level, or there may be one plan that addresses a number of identified risks.

Figures 7.3 and 7.4 refer to a **risk owner**. The risk owner is responsible for ensuring adequate management of the risk, including how and at what intervals a risk will be monitored. If the nature of a risk changes during this process, it may be necessary to revise earlier risk mitigation plans.

## 7.7 Summary

Managing risk is a continuous process. It involves identifying the risks and analyzing them to establish their probability, impact, proximity, exposure and priority. Actions need to be determined and plans produced for these actions, followed by scheduled monitoring and appropriate control. The whole risk management process needs to be made visible by adopting sound communication mechanisms.

Murphy has been quoted as saying, 'whatever can go wrong, will go wrong'. Undoubtedly risks will occur in your project. Most of the risks that you will experience have already been identified in previous projects. Learn the lessons from those. To ensure that those risks have the least impact, one needs to develop a structured risk management approach within your overall project planning.

## Sample questions

1.  Which of the following best defines a contingency action?

(a)  An activity that is planned to take place if a risk materializes.

(b)  An action taken at the start of a project which reduces the potential damage if a certain risk does materialize.

(c)  An agreement that the users accept a particular risk.

(d)  An activity which prevents a risk from materializing.

2.  What is maintenance of a risk log designed to do?

(a)  eliminate risk;

(b)  save money;

(c)  control risk;

(d)  prevent system development failure.

3.  Which of the lists below best identifies what is examined when a risk is assessed?

(a)  Probability, proximity, owner

(b)  Impact, probability, team experience

(c)  Cost, benefit, business case

(d)  Probability, proximity, impact

4.  Which of the following is NOT an action that would mitigate a risk?

(a)  accept it;

(b)  prevent it;

(c)  reduce it;

(d)  transfer it.

## Answers to sample questions

1. (a)    2. (c)    3. (d)    4. (a)

## Pointers to activities

### Activity 7.1

Among the facts that could lead to difficulties in the Canal Dreams scenario are the following:

● Canal Dreams has only recently been created from an amalgamation of six companies.

● Local staff who deal with holiday bookings have little or no IT experience.

● Canal Dreams has no in-house IT development staff which means the organization is going to be dependent on outside suppliers.

● New staff will need to be recruited to run the new booking system at the centre.

## Activity 7.2

| Issue | Possible risk |
|---|---|
| Recent amalgamation | • Local resistance to centrally imposed changes<br>• Differences in procedures at different boatyards |
| Local staff have little or no IT experience | • Resistance because of fear of not being able to cope with new technology<br>• Many errors initially because of inexperience |
| No in-house IT development experience | • Poor quality software (e.g. poor maintainability) because of lack of expertise in evaluating deliverables |
| New booking staff at centre | • Initial mistakes or inefficiencies because of inexperience<br>• Customers receiving poor advice due to lack of local knowledge |

## Activity 7.3

| Issue | Mitigating actions |
|---|---|
| Local resistance to centrally imposed changes | • Take a participatory approach to requirements gathering and system design<br>• Engage in company-wide team-building<br>• Make bookings locally as well as centrally – at least at first |
| Differences in procedures at different boatyards | • Take a participatory approach<br>• Appoint a central user representative with authority to resolve disagreements between users |
| Resistance because of fear of not being able to cope with new technology | • Provide training courses<br>• Visit companies with similar systems |
| Many errors initially because of inexperience | • Provide training courses<br>• Visit companies with similar systems<br>• Use prototypes to give users early experience and understanding of the application<br>• Involve users in testing<br>• Start operations at a quiet period to allow gradual build-up<br>• Provide generous initial user support |
| Poor quality software because of lack of expertise in evaluating deliverables | • Use consultants as quality auditors of products<br>• Employ an experienced IT professional to oversee the project |
| Initial mistakes or inefficiencies because of inexperience | • Involve users in testing<br>• Start operations at a quiet period to allow gradual build-up<br>• Provide generous initial user support<br>• Train existing booking staff<br>• Allow new recruits to spend time working on old system (this may be problematic) |
| Customers receiving poor advice due to lack of local knowledge | • Make bookings locally as well as centrally – at least at first<br>• Train existing booking staff<br>• Allow new recruits to spend time working on old system (this may be problematic) |

# CHAPTER EIGHT

# Project organization

## LEARNING OUTCOMES

When you have completed this chapter, you should be able to demonstrate an understanding of the following:

▶ relationship between programmes and projects;

▶ identification of stakeholders and their concerns;

▶ the project sponsor;

▶ establishment of the project authority (e.g. project board, steering committee, etc.);

▶ membership of project board/steering committee;

▶ roles and responsibilities of the project manager, stage manager and team leader;

▶ desirable characteristics of a project manager;

▶ role of the project support office;

▶ the project team and matrix management;

▶ reporting structures and responsibilities;

▶ management styles and communication;

▶ team building and dynamics.

## 8.1 Introduction

This book endeavours to describe what could be called 'best practice' in project management. Unfortunately, not all organizations follow these good practices, which is one of the reasons why there are so many project failures. In this chapter, we describe the elements of the project management structure that should exist in an organization that is planning and executing a project. In many cases, these roles will be known by different names to the ones we use. Many of the terms we use are loosely based on those used by PRINCE 2, the project management standard sponsored by the UK government. Note that we will use capitalized initial letters, for example Senior User, where this refers to a very specific PRINCE 2 concept.

## 8.2 Programmes and projects

A **programme** is a collection or group of related projects. Historically, IT projects were often treated as individual and separate undertakings with their own distinct aims and objectives. It became apparent that this was a flawed approach and that there were many advantages to be gained by grouping projects together into a programme. These include the following:

- **Reduction of duplication** – two projects may be developing the same or a similar product.

- **Co-ordination of resources** – projects inevitably place demands on an organization's resources and it is far better to coordinate than compete for them.

- **Management of interdependencies** – if projects are dependent on one another they have to be properly co-ordinated. In such circumstances, a change in one project may introduce a delay or a consequential change in another.

Typically each project is assigned a project manager, whereas a programme is managed by a **programme manager** or **programme director**. The programme manager is responsible for the co-ordination activities described above. (Do not confuse programme with **program,** which means a piece of code or software.)

In the Canal Dreams project scenario, the booking application may be one of a portfolio of projects. An associated project may relate to keeping an inventory of the boats and a record of their maintenance schedules. The booking system may need to access the maintenance records in order to check whether a boat is available for booking. A third project may be designing a website that allows customers to book boats directly.

## 8.3 Identifying stakeholders and their concerns

A **stakeholder** is defined as anyone with a valid interest in or affected by an IT project or the products delivered by it. This group includes:

- all project personnel including managers, designers and developers;

- users, sponsors and other members of the organization affected by the project;

- suppliers of software, hardware, consultancy, etc. to the project;

- contractors and subcontractors;

- members of the business community and, of course, financial backers.

Part of the project initiation process is to establish who these people are and to identify their needs and concerns. Processes will have to be set up to ensure that their interests are represented and that they are consulted and kept informed.

**List the main stakeholders in the Canal Dreams booking system implementation project.**

*Activity 8.1*

## 8.4 The organizational framework

In order to manage a project successfully it is essential that a formal management structure is in place.

- Named personnel should be allocated to the roles – but several people may share a role and a single person may have more than one role.

- Roles must carry appropriate authority and responsibility.

- Individuals must carry out their roles correctly and willingly and must clearly understand their objectives. Status within the organization is not sufficient qualification. A period of relevant training may well be appropriate.

At the top of the project organization is a **project sponsor**. This person would be a senior person within the organization who would be able to champion the project at an appropriately high and influential level, normally board level. Crucially, the sponsor controls the funds to pay for the project. Below the project sponsor is the following structure:

- project board;
- project manager;
- stage manager;
- project team leader;
- team member.

In parallel to this structure are three other functions:

- project assurance team;
- project support office;
- configuration management team.

All these titles are based on the PRINCE 2 terminology, so whether or not they exist within a particular project organization will depend on the extent to which the organization embraces the PRINCE 2 approach. However the roles that they represent should be present in all projects to a greater or lesser degree, although they may use different names. Where a project is small, for example, the project manager, stage manager and project team leader roles could be carried out by a single person.

Having established who the project sponsor is, it is necessary to establish a body to control the project – the **project authority**. The British Standards Institution's *Guide to project management* (BS 6079) states that the authority for a project lies with the project sponsor. In PRINCE 2, the project board performs this role. It is probably the case that project boards become crucial with larger projects where there are many influential stakeholders. The main point is that it is clear who has the ultimate responsibility for the project and, inevitably, it will be whoever holds the purse strings.

## 8.4.1  Project Board

The concept and existence of a **Project Board** is of great benefit to the project as it provides a forum in which critical issues can be discussed and decisions taken which are outside the remit and competence of the project manager. If PRINCE 2 is not being followed, the same function may be served by a **steering committee** or a **project management board**. The board, or steering committee, normally includes the project sponsor or someone representing the sponsor. In PRINCE 2 terms this person is known as the **Executive** and is joined by a **Senior User** and a **Senior Supplier**, although in certain circumstances the Executive and the Senior User could be the same person.

Members of the board must be able to make decisions for their areas of responsibility without having to refer to higher authority. It is important therefore to appoint the right people to this board. If they are at too low a level they will constantly be referring back to their managers for a decision and this is likely to introduce unacceptable delays. Too high a level can also be a problem in that getting extremely important and busy staff to attend meetings may be difficult, resulting in the board being ineffective or adding delays to the project.

All projects should derive from the organization's business strategy and be designed to meet specific business and corporate objectives. It is therefore the role of the Executive, apart from looking after the money, to ensure that the project:

● stays in line with the corporate objectives;

● meets the business requirements;

● retains its business case (that is, that the benefits continue to outweigh the costs of the project).

As the Executive represents the project sponsor, he or she will have ultimate authority for the project.

The role of the Senior User is to safeguard the interests of the users and make sure that their requirements are met. He or she will be responsible for ensuring that the user requirements have been accurately captured and for signing off products as having met acceptance criteria. This role is particularly important as, with the best will in the world, requirements will change over time. The Senior User must therefore ensure that any proposed changes to requirements are in line with the business needs, can be justified and do not jeopardise the project as a whole. As was seen in Chapter 4, perfectly valid changes can delay the project to such an extent that it will not deliver the expected benefits.

The role of the Senior Supplier includes ensuring that adequate technical resources, both in terms of skilled staff and the software and equipment they need, are available to the project. They should also support the development team and, when necessary, represent their interests as difficulties arise, for example in the face of continual change requests.

This describes the minimum board structure. Other roles may well be appointed. If all or part of the project is contracted to an external supplier there may well be a senior representative from that organization. If there are a number of contractors then more than one supplier representative may be required. However, this could detract from the main purpose of the Project Board and therefore it may be more effective to set up a separate group to manage the external suppliers.

The **quality assurance function** could also have a representative on the board or may report via another member of the board, usually the Executive.

If the project is large or involves many functions, then it may be necessary to have more than one user representative. Similarly if it is geographically dispersed then additional representatives may be required. As in the case of multiple supplier representatives, care must always be exercised to ensure that the board does not get too big and thus become ineffective. Subordinate structures may have to be developed to give a voice to all those who should have one without detracting from the efficient working of the board. This is always a balancing act, but smaller groups are usually more effective when it comes to decision-making.

The infrastructure management side of the business, that is, those who are responsible for the infrastructure on which the delivered IT system will run, may also be represented. This is important if the development staff have different managers to the infrastructure support staff as there could well be a conflict of interest between the two groups. Normally, however infrastructure management would be represented through the Senior Supplier and, where the project is being delivered by external suppliers, the Senior Supplier would probably be drawn from the infrastructure management side of the business.

The Project Board is a decision-making body. It holds the purse-strings through the Executive and therefore has ultimate control of the project. The project manager would normally be appointed by and report to the board. The board establishes the terms of reference and provides the management framework within which the project manager operates. It is the final arbiter as to whether the project has met its objectives or not. In summary, the board initiates the project, controls its execution and eventually closes it down. Among its functions it approves the following items:

- project terms of reference (including the project initiation document);
- business case;
- budget;
- project plans;
- changes to project plan;
- quality plans, control and assurance processes;
- risk assessment and contingency plans;
- major changes to project requirements.

It receives feedback from the project manager on the progress of the project and also from the quality assurance function. Given the feedback, it is then in a position to sign off each stage of the project or other activity as defined in the plan or require that it be reworked. It thus exercises overall control of the project.

## 8.4.2 Project manager

The project manager is pivotal in the organization of the project and has overall responsibility for the day to day management of the project. The role of the project manager is to ensure that the project is delivered on time, within budget and to the specified quality. A daunting task! The project manager produces, perhaps with the help of others, all the various plans for the project (see Chapter 2).

Monitoring progress against the plan and making adjustments as necessary is an ongoing task for the duration of the project. As milestones are reached, progress will be reported both to the board and to the team members (see Chapter 3). The project manager will have been given a set of tolerances for activity completion and costs within which to work. Any deviation in the plans that is likely to take the project outside these should be reported to the board along with recommendations about the actions the project manager feels are necessary either to correct the situation or to mitigate its effect.

As changes come along, as they inevitably will, the project manager has to assess their impact on the project and report to the board. It is the responsibility of the board to decide whether or not changes should be implemented although, as seen in Chapter 4, this responsibility may be delegated, within constraints, to a change control board.

Should a risk materialize, then the project manager will approach the board for permission to put into effect the agreed contingency plan. Logs of both change requests and risks are kept (see Chapters 4 and 7, respectively). The project manager has the responsibility of making sure that they are kept up to date.

Because project managers have to lead their projects they must be able to set clear goals so that those who report to them are aware of what is required. The simple production of plans and schedules does not by itself achieve this.

### 8.4.3    Stage manager

In larger projects, someone may be appointed to the position of a stage manager. This role is similar to that of the project manager but is specific to the particular stage. Detailed plans, work schedules and monitoring then become the responsibility of the stage manager rather than the project manager. This does not mean that the project manager gives up all responsibilities but more that much of the routine work is carried out by someone else, who may have particular technical expertise for dealing with a particular stage. The project manager is still responsible to the board for the progress and success of the project.

### 8.4.4    Team leader

Team leaders are close to the action. A team leader may be responsible for a specialist group of analysts, designers or programmers and hence work on a very specific part of the life cycle. Alternatively, a team leader may lead a mixed team, in which case the responsibilities would encompass the whole of the life cycle. A team leader usually needs technical experience and knowledge. One very specialized area is that of testing where a team becomes expert at the creation of test data based on the requirements specification and at the exercising of the programs to determine whether or not they meet those requirements.

Team leaders have the task of allocating authorized work packages to specific individuals and helping them to complete the activities within the scheduled time scales. They also act as mentors or advisers when necessary. These are the people who will be most aware how of well or badly a particular part of the project is going. They should be able to feed this back to the project manager at an early stage so that action can be initiated to bring the project back on course before it goes seriously adrift.

### 8.4.5    Team member

At the bottom of the pile is the team member. However good the structure may be, without competent team members the project will not succeed. Ideally, project managers should be able to select their teams. This is rarely possible, so the project manager has to know the staff available and assign them to the tasks which most match their capabilities. With larger organizations in which team leaders or stage managers exist, they may undertake this task or share in it with the project manager. Where teams are being created then there is more opportunity to get the balance of skills and personalities right. The matching of skills to tasks and the creation of a coherent team is critical to the success of a project.

### 8.4.6    Project assurance team

This is usually a function rather than an actual team. The project board may fulfil the role if it chooses to do so. However, depending on the size and nature of the project, it is normally delegated to one person or a small group. Whichever way it is done, the tasks are the same; it is just the degree of activity involved which would be different. Members of the project assurance function are appointed by and report directly to the project board and thus are not under the control of the project manager. Their role is primarily quality assurance (see Chapter 5). They are responsible for checking that the quality control activities as specified in the quality plan are carried out, that standards are observed and that procedures are followed. Being independent of the project manager they are able to give an objective view on the quality of the products being delivered and not just the timeliness of their arrival.

If the project is large and complex several people may be involved in this activity, each with their own specialism, such as security. Three specific roles sometimes identified are those of **business assurance co-ordinator**, **technical assurance co-ordinator** and **user assurance co-ordinator**. Their purpose is to assure the integrity of the project from these points of view. The business assurance co-ordinator would, for example, make sure that the IT application under development is compatible with existing business procedures. The technical assurance co-ordinator would ensure that the operational environment is not compromised by the new system.

In the early stages of a project, the team could also be involved in the setting of standards, the creation of procedures, the establishment of the quality review and quality assurance processes and the quality plan. This is particularly useful if the project is venturing into areas of technology that are new to the organization.

**In the Canal Dreams booking system project, the main driving force behind the new computer-based booking system is the managing director. He has given you a contract to manage the project. You are in regular contact with the managers of local boatyards and the director of finance. You also need to communicate regularly with the IT manager at Canal Dreams who looks after the existing IT infrastructure. The software is to be supplied by the XYZ software house and you have had several meetings with one of their account managers. Identify which type of representative on the Project Board would be likely to represent each of the various stakeholder groups in the scenario.**

*Activity 8.2*

## 8.5 Desirable characteristics of a project manager

Although the project manager is just one element in the project management structure, the role is pivotal. IT project managers traditionally have progressed from technical areas such as software development through system design and team leadership to project management. This, however, may be changing as the project management role nowadays often involves managing the external suppliers of services. Thus, while it is useful to have a good technical background it is not essential in order to manage an IT project.

It is more important to have other skills and characteristics. Perhaps the key quality is having good communications skills at all levels. The project manager may need to present to higher management a convincing case for a course of action. Also the stakeholders have to be kept informed and the project teams must be well motivated. If project managers are to gain the respect and confidence of those around them, they have to be effective communicators, good managers and effective organizers.

They need to have skills in:

- leadership;
- motivation;
- planning;
- negotiation (being firm, flexible, and able to compromise where appropriate);
- delegation.

They need to be:

- responsible;
- reliable;
- available (not just for this project but contactable at all reasonable times);
- intelligent;
- sociable (able to mix well);
- approachable (they should be good listeners);
- knowledgeable in the business area for the particular project.

It is important to appreciate that these characteristics do not just arrive with seniority. A potential project manager must possess some of these attributes and will need training in deficient areas before becoming fully effective.

Although this list may seem daunting, the ability of people who seem quite 'ordinary' to become competent project managers through application, self-discipline and appropriate training is remarkable.

## 8.6 Project support office

Within all projects there are many routine clerical activities which have to be performed regularly, effectively and efficiently. Many of them fall within the remit of the project manager. However, given their routine nature and the pressures on project managers, it is usually more effective to delegate these tasks to others. This gives rise to the creation of what has become known as the project support office (PSO).

The organizational structure that has been described above is usually set up for a specific project. This means that the project board and project manager are appointed for a single project. The project support office, on the other hand, may be to a greater or lesser extent, a permanent organization which supports several usually inter-related projects. The precise tasks it undertakes will vary from organization to organization, but the following list is typical of the processes which would form the day-to-day activities of a project support office:

- time recording;
- updating of project plans;
- risk log maintenance;
- issues (change) log maintenance;
- arranging meetings;
- issuing agendas;

- taking minutes;
- chasing actions;
- configuration management.

### 8.6.1    Time recording

In order to control the project, time recording is vital. Collecting and processing the information required is largely mechanical and can be done by clerical staff. This does not mean that the team leader or project manager should be unaware of what is happening but, relieved of the routine work, they can focus on checking that effort expended is consistent with that expected and, if not, take the necessary action.

### 8.6.2    Updating plans

As details of progress are passed to the PSO they can update the plans and highlight any problems to the team leader or project manager. In the event of changes to personnel, the PSO can help to revise the plans so that the project manager is aware of the impact of such changes and can make any modifications necessary or alert the board to potential problems.

### 8.6.3    Maintenance of logs

The general maintenance of logs can also form part of the work of the PSO. Any potential risks which the project is approaching can be drawn to the attention of the project manager via the risk register (see Chapter 7). Risks that have passed can also be noted. Requests for change can be recorded as they arrive and are passed on to the relevant people for action.

### 8.6.4    Arranging meetings

Much time, most of which involves finding dates and times when all the parties are available, can be spent on arranging meetings. This is plainly a waste of a project manager's time so its delegation to the PSO is natural. These days with electronic diaries it should be much easier but can still be quite time-consuming. Having agreed dates and times the PSO can then organize the necessary room and support requirements such as coffee or meals and, if required, overnight accommodation. Most meetings will have a set agenda so the issuing of these again can be routinely delegated to the PSO. It is simple to check with the chair of the meeting to see if any changes are required. Most meetings will also have other documents that will need circulating. The PSO can take on this responsibility, particularly if they carry out the configuration control function as well.

The recording of the minutes of meetings and the subsequent distribution of the minutes is also a straightforward task for the PSO. The key thing with the minutes is to ensure that all actions are identified along with who is responsible for carrying them out and in what time frame.

### 8.6.5    Configuration management

The whole of Chapter 4 has been given over to change control and configuration management. Here it is enough to say that the ability to keep track of documents and products to ensure that everybody is working with the latest version is very important to the success of the project. Like the PSO, within which it may function, configuration management has a life beyond the duration of the project, as the delivered IT system may continue to require amending and updating until it is finally replaced.

## 8.7 Project team

A project team is defined as a small number of people with complementary skills who are committed to a common purpose, performance, goals and approach who are directly or indirectly accountable to the project manager.

> Michael W. Newell and Marina N. Grashina, *The Project Management Question and Answer Book*, American Management Association, 2004.

Project teams work in a different way from operational staff. A project team is brought together for the sole purpose of achieving the project objectives. On completion of the project, the team is disbanded. Team members may be drawn from a **specialist division** within the organization, such as the IT department, to work on the project and may be assigned to other projects when it is over. Others may be drawn from **functional departments** and will go back to their original jobs at the end of the project. The impact of being in a project team will be quite different for the two types of people as will their expectations during and after the project.

It is easy to assume that a project team is going to be located together. This is best but it may not always be possible. If the team is drawn from different departments, they could well be located in different parts of the country. In such circumstances, it may be possible to bring them together for periods at a time. However they could be located in different countries which would make that much more difficult. Part of the solution to these problems could be matrix management.

## 8.8 Matrix management

So far it has been assumed that all staff report to the project manager, either directly or through team leaders or stage managers. This is generally the best way of managing a project but it does not always apply in practice, particularly if team members are drawn from a variety of departments.

Senior staff on a project, that is the stage managers and team leaders where they exist, should always report to the project manager. If this is not the case the exercise of managerial control may be weakened to a point where it would be impossible, with any certainty, to keep to a delivery plan.

In matrix-managed teams, an individual has more than one reporting line. The most common example of matrix management can be found on board a ship. The captain of the ship has overall responsibility for everything that happens on the ship when it is at sea. Its complement of sailors will include navigators, engineers, electricians, cooks, medical staff and others. Each specialized trade will also have a shore-based manager to whom they will report and who is responsible for their performance on the ship and their training when ashore.

In any organization, it is probable that a number of projects would be under way at the same time. Each project requires a number of different types of skill. For example, business analysts may help to help elicit the requirements and ensure that they are clearly understood by the technical members of the project team. Analysts and designers would then be required to implement the requirements. They could well be part of a specialist department within the organization and report to a separate manager.

In today's environment there is likely to be another department which manages all the IT infrastructure. They could second someone to the team to look after their interests and guide the designers so that the proposed solutions are compatible with the existing structures.

The developers (or programmers) may be part of a separate pool of development specialists or even come from a specialist group from outside the organization. The testing team may come under a different department with its own line management. Putting all these together gives rise to a matrix structure as shown in Table 8.1.

**Table 8.1  Typical matrix organization**

| Project | User department | Analysis and design department | Infrastructure management department | System development department | Software testing department |
|---|---|---|---|---|---|
| A | ✓ | ✓ | ✓ | ✓ | ✓ |
| B | ✓ | ✓ | ✓ | ✓ | ✓ |
| C | ✓ | ✓ | ✓ | ✓ | ✓ |

This sort of structure has the advantage that the project manager has a team of specialists and can concentrate on the project in hand and not be concerned with issues such as the long-term development of the staff involved, as this would be the responsibility of the line managers. It does however suffer from the big disadvantage that the line managers may call their staff off the project for some more urgent task. This makes planning and control more difficult. Ideally these issues should be clearly discussed at the beginning of the project and a strategy agreed by the project board to which all line managers sign up.

Even where all the staff, including the project managers, work for the same line manager they could well be working on more than one project. Each person will have their own perceived priorities and put more effort into one project and less into another. This could be because one task is found to be easier or more interesting or more rewarding than another. This can lead to a slippage on one project while the other is progressing to schedule. Even in this situation, staff can still be redeployed from one project to another due to an unforeseen event (a risk that has materialized), for example, a colleague being absent through illness at a critical time.

We noted in Section 8.2 that when there are many projects being carried out at the same time, particularly if they are related, there is often an umbrella organization known as **programme management**. Where resources are limited, which is the normal situation, a **programme board** would have the remit to make the final decisions about the allocation of resources between projects. While this may be irksome for the individual project manager, it is for the benefit of the organization as a whole. Obviously plans have to be revised to take account of this situation and project managers cannot be held responsible for delays to their projects due to such changes.

With matrix management, the level of control that the project manager will have over the project team will vary. Where an individual is seconded to the project for its duration then the project manager has greater control – as with the ship's crew. This is in many ways the ideal, but it does have its drawbacks. With every project, the level of resource required will vary over time. For example in the early, analysis, stages of a project only one designer may be required and that only part-time. However as the analysis phase is completed, more designers may be required. The reverse is true of analysts. Thus if these staff were permanently with the project they would be under-utilized at times, adding an unnecessary cost to the project.

**List the types of people (including external suppliers) required by the project to implement the Canal Dreams booking system project.**        *Activity 8.3*

## 8.9  Team building

A team in an IT project is usually a collection of specialists with a requirement to work together towards a common goal. It is most usual that the composition of the team is based on a great many compromises and is never the ideal combination of skills and personal characteristics. Making the group of individuals into an effective team is an interesting and rewarding management challenge. There are many common-sense aspects and a number of theoretical approaches to developing an effective team.

The 'Tuckman' model describes four basic phases through which a team goes before it becomes fully effective:

● **Forming.** The group members have just been brought together and are probably hesitant about their new environment, unsure of their new colleagues and possibly nervous about future developments. Members are polite to one another, tend to accept authority and tread carefully. Some initial contact with colleagues reveals common ground and possible allegiances.

● **Storming.** Individuals within the group have started to assert themselves and to form alliances. Some conflict may arise as a 'pecking order' becomes established. Aims and objectives are becoming clearer but there are different views on how to proceed with the tasks ahead. Members now have a sense of belonging to a team, are gaining confidence and are likely to challenge both authority and the validity and feasibility of the tasks ahead.

● **Norming.** Internal conflicts are hopefully resolved and the team members feel more comfortable and relaxed with their colleagues and their new surroundings. An acceptance of common values and behaviours develops, with open communication and constructive cooperation. The team is working as it should with its overall capability being greater than the sum of its parts.

● **Performing.** The team is fully functional and has become a cohesive unit. Team morale is high with good cooperation between members and a shared responsibility for the common goal. Team members are working hard and getting satisfaction as the team achieves its goals.

A fifth stage, **adjourning**, is sometimes added to describe when the team breaks up at the end of the project. The Tuckman sequence of phases clearly represents an ideal situation in ideal circumstances. Good team leadership and people management are essential to allow the team to progress through the phases and indeed the final stage may never be achieved if the personnel or the project circumstances are not right. It is quite possible to slip back a stage or two if unexpected developments are not well-managed, for example:

● changes in team personnel – new arrivals can disrupt team morale and stability;

● a change in direction of the project may mean much team effort has been wasted;

● a change of leadership may mean the team needs to adapt to a new style and could revert to the storming stage.

## 8.10 Team dynamics

Building a team involves finding people with the appropriate skills who are available when you need them and are motivated to perform the tasks required of them. However, if the team are to work well together, a satisfactory mix of personality types and personal attributes is essential. A system for analyzing and categorizing people's personal characteristics was developed by Belbin, who defined nine **team roles**:

● **Shaper** – an energetic team member with a strong need for achievement who drives the team along;

● **Plant** – a creative and innovative team member (the term 'plant' is used because it was found that planting such a person in an uninspiring team was a good way to improve its performance);

● **Resource investigator** – a team member who makes contacts outside the group to bring in ideas and information and to acquire materials/resources;

● **Co-ordinator** – a chairperson who promotes decision-making and delegates well (not necessarily the team leader);

● **Monitor evaluator** – a team member who is analytical, able to assess ideas and options but is not creative;

● **Team worker** – a team member who helps to maintain team spirit and cohesion.

● **Completer finisher** – a conscientious and painstaking team member who is concerned to get things finished (this is a very important team trait);

● **Implementor** – a team member who attends to details, is hard-working and organizes the practical side of the project;

● **Technical specialist** – someone who can provide the team with technical expertise.

(For further details, see R.M. Belbin's *Team roles at work: a strategy for human resources management*, Butterworth-Heinemann 1993.)

It is not suggested that a team has to have one person of each type or that each team member falls into one role. An individual can have attributes from a number of different roles. The idea is that a team needs a satisfactory mix of roles within its members to perform well. A Belbin analysis may help to define a team weakness or indeed to clarify the reasons why a team is not performing well or why conflicts keep arising in an otherwise skilled, competent team.

## 8.11  Management styles

There are as many styles of management as there are managers. The different styles and approaches depend as much on the personality and capability of the manager (or leader), as on the environment and the prevailing circumstances. The manager may well have a natural preference for a certain style but will have to vary the approach depending on the circumstances in order to be successful.

For example, a **task-orientated** leadership style is sometimes distinguished from a **relationship-orientated** leadership style. Task orientation focuses on the technical aspects of the work, while relationship orientation emphasizes such things as individual motivation and team morale. When a team is developing – the forming stage – it will need more direction than when it is fully functioning – in the performing phase. A task-orientated approach would be more effective than a relationship-orientated approach because the team members are not yet familiar enough with the tasks to be carried out and the environment in which they are to be carried out.

This distinction between task orientation and relationship orientation overlaps with that between **autocratic** and **democratic** styles of management. When staff are new to a project a more autocratic approach may be called for as staff are not yet sufficiently knowledgeable to be involved in complex decision-making. Individual team members will react differently to these two approaches. Some people prefer to have clear direction and to leave decision-making to others, whereas other team members like to feel they have a part to play in establishing the direction of the project and wish to provide an input to the decision-making process. An intermediate approach is **consultative**.

- Autocratic management provides clear direction and quick decisions. The leader is seen to be decisive, firm and effective. However, it can be synonymous with an uncaring, remote, unapproachable, controlling or bullying management style. An autocratic leader can demotivate a team that is working in a climate of fear (members may seek an alternative project).

- Democratic management shares responsibility for decision-making and for the team's performance: thus the team is likely to be more committed to the project. This style can be perceived as weak and management can be seen as avoiding responsibility for difficult decisions. It may be difficult to enforce discipline if conflicts develop.

- Consultative management seeks the opinions and views of the team prior to a decision being made. Although team members are consulted the final course of action may or may not reflect their views. It is seen as a compromise between autocratic and democratic management with the benefits, and possibly the pitfalls, of both.

A politically acceptable and effective manager would need to stay away from extremes and be able to adapt styles of management to the situation and the people involved. Ideally such a manager could be decisive and effective where a quick decision was required and approachable and consultative on other occasions and thus gain commitment from the team where joint decisions and consensus were appropriate.

Whichever approach is used, the project manager must be an accomplished communicator and have the ability to use the most appropriate **methods of communication** in order to be fully effective.

## 8.12 Communication methods

Methods of communication for IT projects can include:

- memos;
- newsletters;
- meetings;
- presentations;
- progress reports;
- telephone conversations;
- text messages;
- email messages;
- letters;
- drawings;
- one-to-one conversations;
- video conferencing;
- intranets and extranets.

Methods can be categorized as **active** or **passive** and **formal** or **informal. Active** methods, such as a telephone conversation, require a response or reaction so that there is reinforcement or confirmation that the information or message has been received and understood. **Passive** methods such as a newsletter have no such confirmation. It is left to chance whether or not anyone reads or understands the material and should not be used for important messages.

**Formal** methods of communication have a set structure, such as a meeting of a project board, in contrast to **informal** methods such as a conversation, which carries no particular format and is not usually recorded.

It is also possible to categorize methods depending on whether the participants have to be in the same time or available in the same place for communication to take place.

**Categorize each of the methods listed above as same time/same place, same time/different place, different time/same place, or different time/different place.**

*Activity 8.4*

By categorizing methods in this way it is possible to assess the suitability of a particular method. It is best to put together a communications plan setting out the methods of communication that should be used during the project for each set of circumstances and exactly how they should be applied. For example, you may decide that a short weekly meeting (an active, formal method) with the project sponsor would be better than an email.

By setting out a detailed **communications plan** with all stakeholders, rather than leaving things to chance, you stand a much better chance of getting it right.

# Sample questions

1. Which of the following is not a name given to the group which has responsibility for committing resources to the project and approving variations to the project's objectives?

(a) project board;

(b) project management board;

(c) project support office;

(d) steering committee.

2. Which of the following terms is used to describe an organizational structure where staff are responsible to a project manager for the duration of a project but also have a manager who is responsible for their long-term staff development and work programme?

(a) the Tuckman model;

(b) configuration management;

(c) matrix management;

(d) task orientation/relationship orientation.

3. At which stage of team formation does a team become fully functional as a cohesive group?

(a) performing;

(b) norming;

(c) storming;

(d) forming.

4. Which of the following is an example of different time/different place communication?

(a) a project board meeting;

(b) a checkpoint report;

(c) a telephone conversation discussing a problem a user has with an IT system;

(d) video conferencing with the supplier of a software application.

# Answers to sample questions

1. (c)   2. (c)   3. (a)   4. (b)

# Pointers to activities

## Activity 8.1

Stakeholders in the Canal Dreams project include existing IT support staff, customers, boatyard staff, booking staff, the managing director, the central finance department, the personnel department, the premises manager (who has to find room at head office for a booking centre), IT equipment suppliers and software suppliers.

## Activity 8.2

The Senior User represents the interests of boatyard staff, booking staff and the central finance department.

The Senior Supplier reflects the views of existing IT support staff, IT equipment suppliers and software suppliers.

The Executive represents the interests of the managing director.

It could be argued that the personnel department, who will have to recruit new booking staff and perhaps relocate existing staff as a result of this project, and the premises manager, who will have to find accommodation for new staff, are suppliers of services to the project.

Who will represent the interests of the customers of Canal Dreams? It may be the marketing department, who may have conducted market research into the preferences of potential customers, or the booking and boatyard staff, who have had most contact with customers in the past. These would be represented on the project board by the Senior User.

Note that terms like 'Senior User' simply represent roles. Each project must form a project board with representation that is sensible for that project.

## Activity 8.3

- Users;
- Analysts;
- Designers;
- Database designers;
- Network/telecommunications specialists;
- Hardware specialists;
- Software developers;
- Testers;
- Trainers.

## Activity 8.4

|  | Same place | Different place |
|---|---|---|
| same time | meetings<br>presentations (normally)<br>one-to-one conversations | telephone conversations<br>video conferencing |
| different time |  | memos<br>newsletters<br>progress reports<br>text messages<br>email messages<br>letters<br>drawings<br>intranets and extranets |

# About ISEB

Through the Information Systems Examinations Board (ISEB), the British Computer Society provides industry-recognized qualifications that measure competence, ability and performance in many areas of IS, with the aim of raising industry standards, promoting career development and providing a competitive edge for employers. ISEB qualifications add value to a professional career by providing both the means and the platform for recognition and enhanced career development. As well as being universally accepted, ISEB qualifications are carefully tailored to meet the needs of IS practitioners at all stages in their careers. The qualifications are designed to be both accessible and relevant. They cover all major areas of IS including Management, Development, Service Delivery and Quality. Thanks to close links with industry representatives and subject sponsors, they also encompass the latest technological advances and new methodologies.

## Qualification areas

ISEB offers exams in the following areas:

- Business and Management Skills
- Business Systems Development
- Data Protection
- Dynamic Systems Development Method (DSDM)
- Freedom of Information
- Information Security Management
- IS Consultancy Practice

- ITIL Infrastructure Management
- IT Service Management
- Programme and Project Support Office (PPSO)
- Project Management – Certificate
- Project Management – Foundation Certificate
- Software Testing

## Benefits of ISEB

ISEB qualifications offer a range of benefits to both the employer and the individual. Employers gain from improved IT performance, improved workplace motivation and attraction of high calibre staff, whilst maintaining independence from any particular package or software provider. As well as external and international recognition of their ability, individuals benefit from high quality relevant training, increased job satisfaction and an individual sense of achievement.

For general information about ISEB and other BCS professional development initiatives, please contact:

The British Computer Society – ISEB Qualifications

4th Floor, Minton Place, Swindon SN1 1AB

Telephone: +44 (0) 1793 417542

Fax: +44 (0) 1793 480270

Email: isebenq@hq.bcs.org.uk

Website: www.iseb.org.uk

# Index